Science and Hope

SCIENCE AND HOPE: A FOREST HISTORY

John Dargavel and Elisabeth Johann

First published 2013 by
The White Horse Press, 10 High Street, Knapwell, Cambridge, CB23 4NR, UK

Set in 11 point Adobe Caslon Pro and Lucida Sans
Printed by Lightning Source

British Library Cataloguing in Publication Data
A catalogue record for this book is available from the British Library

ISBN 978-1-874267-73-7

TABLE OF CONTENTS

ACKNOWLEDGEMENTS vii

COMMENCING 1

FOUNDATION 11

1. *Measuring* 15
2. *Tending* 27
3. *Profiting* 40
4. *Regulating* 52

EXTENSION 63

5. *Introducing* 65
6. *Converting* 76
7. *Conquering* 93

DEVELOPMENT 107

8. *Breeding* 109
9. *Fertilising* 123
10. *Planning* 133

DIVERGENCE 147

11 *Balancing* 149

12 *Excising* 168

13 *Devolving* 183

MILLENNIUM 201

14 *Warming* 203

15 *Sustaining* 218

CONSIDERING 229

APPENDIX: List of species 235

BIBLIOGRAPHY 239

INDEX 257

ACKNOWLEDGEMENTS

We salute all those librarians and archivists whose work has conserved the science of the past: their unseen presence pervades this book. Some of them in the Internet Archive of Canadian Libraries (Project Gutenberg), the University of Michigan, and the University of Toronto have given their books digital lives that we have appreciated. We thank the archivists of the University of Maine, who provided information, and the President and Council of the Royal Society of London, who allowed access to their *Collected Papers*. The book was written while one of us (JD) was an honorary Visiting Fellow in the Fenner School of Environment and Society in the Australian National University. He expresses his thanks to the Director, Professor Stephen Dovers, and the professional staff of the School for their unstinting assistance. We have been fortunate in having Clive Hilliker prepare the maps and figures and to have had help and advice from Max Bourke AM, Alan Brown, Robin Cromer, Ricki Dargavel, Eva van Gorsel, Mark Greenaway, Mary Hobley, Peter Kanowski, Sebastião Kengen, Jenny Mills, Greta Olsen, Jan Oosthoek, Mike Roche, David Salt, Brian Turner and Roger Underwood: we thank them all.

COMMENCING

We tell the story of the hopeful science and trusting art of forestry. Ours is no tale of triumph; the outlook for the world's forests is too bleak for that: while many forests are flourishing, the climate is changing, tropical forests are disappearing, others are degrading, species are being lost, governments are dithering and international conferences are failing. It would be easy to deplore the grind of the world system, but it would be a partial story. Ours is another, longer story; one of inquiry, of science and persistent endeavour to find a better future for the forests. Ours too is a partial story; science is not enough, but it is necessary.

Foresters have tried to know their forests scientifically for over three centuries and have hoped to apply their knowledge to good effect, even though they could not live to see the sylvan futures they envisioned. Theirs is a long-term business: fifty, one hundred, two hundred years or more; in perpetuity they dream. Only in the coppices of the Old World or in the plantations of the New World might seven, a dozen or thirty years suffice.[1] Forestry is not an affair of their dreams alone. Toil as they may or not, it is embedded deep within the political economy and ideas of its time. It is complicated too because forests are the most intricate of all natural resources; their web of life, from the highest branch to the deepest root provides a cornucopia of goods and values that people clamour for. Their political power pervades the forests in ways that variously shape the foresters' endeavours. Deforestation is their ultimate defeat, but foresters hope that its advance can be arrested and they strive to restore forests degraded in the past.

The science and ideas of forestry are the focus of this book. Much has been written on the origins and development of modern forestry in various countries, and on the people and institutions involved, but there is little in the forest history literature that explains what the science actually *is*. To many that is the domain of textbooks. We seek to bridge the gap between the histories and the textbooks.

Trees so dominate the forest's web of life that planting, tending or cutting them down are the core activities. It is only trying to grapple with the long-term consequences that make the science and decisions difficult. The ideas of how foresters made their decisions are considered in this book. Many other activities are involved in managing forests – building roads, providing recreation, managing game, controlling grazing, safeguarding all the varieties of plants and animals and guarding the water rippling in the forest steams. There is a horde of specialities within forest science, from root physiology to wood science, to foliage disease, to forest engineering, for example, each with its own history. We concentrate on the core concerns of managing forests to achieve some vision for their future.

Forest science in the sense of forest knowledge has an ancient history documented since classical times and applied within the intricate social and legal systems of medieval Europe.[2] We are concerned with the modern form that was founded in Europe early in the nineteenth century. It needs some introduction here because it emerged only slowly from the centuries of practice and knowledge. It arrived unevenly and rural practices were slow to change. European societies were still in the age of wood, as they had been since the Bronze Age.[3] Wood was the fuel, the shelter, the ships, the fences, the items of everyday life. The forests were its source, but also provided forage, bedding for animals, clean water, food and medicine; they were integral to the rural economies. The growing urban and industrial economies were demanding more and more wood; they depended on it, and the forests had to provide it, not only for industrial energy, but also for chemicals, like potash for the glassworks. The regimes for managing the forests that could be traced to the ancient world and had flourished with intricate detail in the Middle Ages were disrupted. New ways had to be found.

They could not be entirely new; there was knowledge bound into indigenous custom, accumulated by rural practice and written in the ancient Mediterranean world to draw on.[4] For example, in the third century BC Theophrastus had described the characteristics of many tree species in his *Historia Plantarum* and in the first century AD Pliny the Elder, had described in his *Historia naturalis* how to cultivate chestnut, oak, cypress and pine by sowing their seed. At the start of the fourteenth century, an Italian, Petrus de Crescentiis had drawn on classical and medieval sources to write his *Liber Ruralium Commondorum*. He included treatises on growing trees, silviculture and managing forests, distinguishing between those that were natural and those derived from human intervention.[5] Forests were so important that his Latin manuscript was translated into Italian, French and German, in which language it was the first book to be published after the bible.

During the seventeenth and eighteenth centuries, it was mathematics, astronomy, natural philosophy – much of which we now call physics – and chemistry that formed the exemplars of quantitative science, while natural history – in botany and zoology – developed the classifications of living things. The rise of science, an enlightened belief in rationality, the industrial revolution, political revolutions, the spread of democracy, agrarian change, urbanisation, the ascent of capitalism, the exploration of the New World: all these were in the packet of changes that created the modern world system.

Modern science and forestry came together on Tuesday, 16 February 1664 in a moment that can mark the commencement of this story and the threads that run though it. This was the day on which John Evelyn presented his book, *Sylva or a discourse on forest trees and the propagation of timber in His Majesties Dominions to the Royal Society*, the first lasting scientific society to be established in Europe.[6] *Sylva* was the first book that it published and the most enduring.[7] Nothing before on forestry had ever been as thorough or had the imprimatur of both science and the state.

Evelyn confronted a national problem. 'Truly, the waste, and destruction of our Woods, has been so universal', he wrote, 'that I conceive nothing less than an universal Plantation of all the sorts of Trees will supply, and well encounter the defect'. Deforestation – ' the disproportionate spread of Tillage' – during the Civil Wars and increasing demands for wood meant that the forests had to be restored.[8]

The scarcity of timber for the navy was the urgent problem. In September 1662 the Commissioners of the Navy asked the newly-chartered Royal Society for its advice.[9] In today's parlance, we would say that it engaged the Society as a consultant with a five-point brief: could the run-down royal forests be restored by planting; how much should be planted and how; should felling on private land be regulated by the state; should house construction in London be regulated to save using oak; and should landowners be required to plant a proportion of their land with oak or elm trees? The Commissioners' 'quaeries' were worded differently and any planting had to be consistent with 'his Maj^{ty} benefit and pleasure in his Game ... Forests, Chases and Parks'. Only forests within thirty kilometres of the sea or a navigable river needed to be considered, as the cost of hauling timber any further on land was prohibitive. Clearly, science, public policy, political power, different uses of the forests and the mundane realities of production were connected in the Commissioners' minds.

Evelyn's vision for restoring the royal forests and estate woods mirrored a national yearning to restore order again after the years of war and, with Charles II restored to the throne, it reflected his royalist politics. It

John Evelyn (1620–1706)[10]

John Evelyn was the second son of an English family in the upper strata of the gentry. His father, Richard, became Sheriff of Surrey and Sussex and had a well-wooded estate at Wotton in Surrey where John Evelyn was born in 1620. From the age of five, he was brought up by his grandmother at Lewes in Sussex. As a young man, he studied at the Middle Temple and Balliol College, Oxford. He loved Wotton and designed its gardens. It passed to his elder brother George when his father died in 1640 and eventually to Evelyn who spent the last seven years of his long life there.

John Evelyn, holding *Sylva*

By Thomas Bragg, after Sir Geoffrey Kneller, Bt. line engraving 1818, National Portrait Gallery, London, D2317.

He lived through the turbulent times of the Civil War (1642–1646), the Commonwealth Government led by Oliver Cromwell, the Restoration of Charles II (1660), three wars with the Dutch and war with France, the Glorious Revolution of 1688 that deposed James II, the plague in 1665 and the Great Fire of London in 1666. He spent four years (1643–1647) in voluntary exile on a grand tour to France and Italy, where he visited many of the great gardens. In 1647 when he was 26, he married the twelve or thirteen year old Mary Browne in Paris. She was the daughter Sir Richard Browne, then serving at the court in exile of the future Charles II. They did not live together until four years later when he had returned to England and acquired a small estate, Sayes Court, at Deptford on the Thames estuary, east of London. They had eight children, four of whom died in infancy. He was survived only by his wife Mary and their daughter, Susanna.

Evelyn was an active member of the Royal Society from its start in 1662 and he held several public appointments, including being a commissioner to care for those sick and wounded in the Dutch wars, and later to build Greenwich hospital for retired naval sailors. Although he wrote and published on several topics, his great interest lay in horticulture, arboriculture and garden design. He developed the gardens and plantations of Sayes Court, advised other landowners and assembled a mass of information for an unpublished encyclopaedia of gardening, *Elysium Britannicum*. His *Sylva* was highly regarded in his lifetime and in the eighteenth century. Evelyn is best known today for his lengthy *Diary* which, like that of his friend Samuel Pepys, gives a valuable insight to life and manners in a period of great change.

also reflected his interests in landscape design, as *Sylva* was as much about aesthetic values as about naval timber and other products. For Evelyn, beauty and utility went hand-in-hand. Having imagined this sylvan future, he set out the information to accomplish it. Evelyn, a tree enthusiast, saw planting as the solution. The most useful species and the best ways of growing them had to be found. *Sylva* was a practical text book.

Evelyn's *Sylva* epitomises the era when science was emerging as a form of inquiry based more on observation than on obeisance to the classical texts. Although it is replete with references to the Greek and Roman classics, this was a literary convention to be followed by any writer who wished to be taken seriously.[11] He gained his knowledge from the traditional knowledge of 'skilful gardeners' across the country and on his own estate at Sayes Court, at Deptford, east of London.[12] His science was thus a mixture of making his own simple experiments at home and learning about the best English traditional practices of his day. The change can be seen in his nursery practices. With the required nod to the classics, he noted that Pliny would keep a seedling in the nursery for two to three years, much as he recommended, but that Cato was quite misguided in saying that they should be kept until they were much larger. Evelyn was considered an 'ingenious gentleman' in his day, as the term 'scientist' did not come into general use until the 1840s.

France had the same problems, but larger forests. The accessible ones had been degraded over the centuries by incursions, theft and damage. Improving their management was the first priority; planting was less important. The multitude of customary and legislated rights that had been accumulated since the Middle Ages made reform difficult, the national administrative structure, *Eaux et Forêts*, had deteriorated and the state's forest income had fallen drastically.[13] At almost the same time that the Admiralty Commissioners were writing to the Royal Society in London, the royal administration in Paris started to tackle France's forest problem.

Jean-Baptiste Colbert, a powerful minister with increasing responsibilities for finance, the navy and commerce drove reform. A survey of the nation's forests in 1661 documented widespread corruption, the plethora of ancient rights and the lack of systematic management. It was followed in 1669 by a national Ordinance that asserted the state's power over all the royal, church, communal and private forests, and specified how they were to be managed.[14] It kept the principle of common rights, but codified them and extinguished any that were less than a century old.

The Royal Academy of Sciences, started in 1666 on Colbert's initiative, became the centre of French scientific inquiry, advice and publication. The Royal Society in London developed links with the French Academy and other scientific societies by electing foreign members. For example, two multi-talented Frenchmen – Henri-Louise Duhamel du Monceau, hailed as the 'father of French silviculture', and the great encyclopaedist, Georges-Louis Leclerc, Conte de Buffon, who had collaborated with Duhamel in forest experiments – became members of both societies. With exchanges of publications and people, forest science became an international inquiry.[15]

By early in the eighteenth century, the great task of restoring the forests had been set in motion. In England, Evelyn's vision of doing so by planting was virtually ignored on the Crown's land, but the estate owners took up planting during his lifetime and throughout the eighteenth century, using *Sylva* as their textbook.[16] By early in the nineteenth century plantations had evolved from being a patriotic adornment to gentlemen's estates to being an economic practice, particularly with conifers in Scotland and northern England. Throughout the eighteenth and early nineteenth centuries, British forestry stayed at the level of what we now call practical silviculture, albeit of a high standard. In France, the national vision of restoring all the forests by better management and control was put firmly in place, only to be grossly disrupted during the Seven Years War (1756–1763) and the Revolution of 1789. The task of restoration there had to be taken up again in the nineteenth century.

'Woods' as well as 'forests' are included, although their meaning has varied with time and place. In Evelyn's time many European forests were still royal domains, whereas woods were smaller areas in the agricultural landscape managed by their villages or owners. Now, forests are categorised more by their vegetation and are defined by their height and the proportion of ground they shade, while woodlands, so common in semi-arid zones of the world, are defined by their sparser, shorter trees.[17]

The book is organised in a roughly chronological sequence. The first section of four chapters – 'Foundation' – describes the 'classical' phase of forest science as it evolved in Europe from the end of the eighteenth century. Its chapters deal with measurement, silviculture, economics and regulation.

The second section – 'Extension'–deals how forestry spread during the mid-nineteenth to the mid-twentieth centuries by introducing trees from other parts of the world, by converting mixed forests to monocultures and in an imperial form across Europe's empires.

Most forest histories have concentrated on these foundation and imperial phases, but this book proceeds to examine the hopes and science of forestry since the 1950s. The third section – 'Development' – deals with the scientific advances in genetics, fertilisation and planning that underlay the expansion of intensively managed industrial plantations that now provide 35 per cent of the world's roundwood.

The fourth section – 'Divergence' – is devoted to the hopes and science of forestry during the fraught period from the 1970s, during which three distinct futures for the natural forests were envisioned. One adapted past forms to take a more environmentally sensitive, ecologically-based practice under the rubric of 'multiple use' forestry. In contrast to this was a focus on biodiversity and conservation science that was often associated with a rejection of human use and an expansion of exclusive protected areas. The third strand focused on human use at the village level in developing countries, under the rubric of 'social' or 'participatory' forestry.

The final fifth section – 'Millennium' – deals with the twenty-first century, in which the Earth's changing climate became an overwhelming issue for politics and science. The amount of forest science being undertaken had steadily increased over the centuries, but climate change demanded that the forest scientists measure and model the forests' influence on the atmospheric carbon cycle. It added to the concerns about deforestation and the loss of biodiversity, but also offered a vision of how the forests might be able to diminish the change. In contrast to this global dimension, the following chapter focuses on the local and particular, for it is there that the forests have to be sustained. It describes how traditional knowledge of the past was investigated to see if it offered guides for the uncertain future; how the cultural richness of landscapes might be maintained; and the drive for sustainable forest management practices.

The book ends by reflecting on the course of the three centuries of forest science: hopeful work that is always in progress in a daunting world.

Notes

1. Coppices are areas of woodland where the trees are cut periodically and the shoots from the stumps are allowed to grow, typically for 7 to 20 years until they are cut and the process repeated.
2. For the ancient world see Hughes 1996, 1997. For medieval forest practice and control in France see Bechmann 1990; in Germany see papers in Vavra 2008; in England see James 1981 and Cardigan 1949.
3. Perlin 1989.
4. Hughes 1994, 1997 reviews how the ancient Mediterranean world thought about forest and other environmental problems.
5. Mantel 1980.
6. National scientific societies were founded in Italy (1657–1667), England (1662), France (1666), Prussia (1700) and other countries later.
7. Evelyn [1664] 1995. Evelyn prepared five editions of *Sylva* in 1664, 1670, 1679, 1706 and 1729, the last appearing a few months after his death. Alexander Hunter prepared five revised and enlarged editions in 1776, 1786, 1801, 1812 and 1825 and an abridged edition was prepared by Mitchell in 1827. Eight further editions of historical and antiquarian interest were produced in the 20th and 21st centuries.
8. Evelyn selected 17 trees as being suitable for plantations: Alder, Ash, Beech, Birch, Chestnut, Elm, Hazel, Hornbeam, Lime, Maple, Oak, Poplar, Rowan, Sycamore, Walnut, Whitebeam, Willow.
9. The Society was founded in 1660 and granted its royal charter on 15 July 1662, Hunter 1989. 'Questions from the Principal Officers and Commissioners of the Navy brought to the Royal Society by Sir R. Moray on 17 September 1662, Quaeries touching ye Preserving of Timber now growing, And planting more in his Maj. Dominions of England and Wales', Archives of the Royal Society, 'Classified Papers', Cl.P, vol. 10(3): 20.
10. Darley 2006 is the most recent of several biographies.
11. Evelyn referred to Cato, Juvenal, Renatus Rapinus, Lucretius, Pliny, Theophrastus and Virgil, for example.
12. Archives of the Royal Society, 'Classified Papers', Cl.P, vol. 10(3): 65. He was a member of the Royal Society's Committee on Agriculture that was intent on 'composing a good History of Agriculture and of Gardening, in order to improve ye practise thereof'. Its correspondence network enabled Evelyn to learn from the practices of 'skilful gardeners' across the country. He tried different ways of growing trees, vegetables and ornamental plants on his own estate.
13. Bechmann 1990.
14. See Reed 1954 for a history of the French forests. Brown 1883 gives an English translation of the Ordinance and discusses the condition of the forests. In steps towards restoring the forests, it specified that the church and communes (local governments responsible for common land) were to keep a quarter of their land under forest and that the boundaries of the royal, church and communal forests were to be surveyed and permanently marked to prevent incursion by farmers. It specified how the forests were to be managed so that the coppice areas were to grow some of their trees for timber, even on the private land. The coppice was to be cut on a ten-year cycle. Under a 'coppice with standards' system, 16 trees per arpent (47 per hectare) were to be allowed to

grow on and all timber-sized trees were to be kept until they were sold, as prescribed under the regulations. Silvicultural systems are discussed in Chapter 2.

15. In 1735 and 1740 respectively, Hartley 2010. See also Bonnaire 2000; Fellows and Milliken 1972.
16. The Royal Society commissioned a fifth edition of *Sylva* in 1729 and a member of the Royal Society, Alexander Hunter, revised five further editions, published between 1776 and 1825. As Anon. 1913 remarked on Hunter's 1812 edition, 'no man will sit down to the text of the *Sylva* as a book of science ... an experienced nurseryman in partnership with a tolerable botanist, would produce a better guide for a modern planter'.
17. The United Nations Food and Agriculture Organisation defines forest as being land with tree crown cover of more than 10 per cent on which the trees should be able to reach a minimum height of 5 metres. It defines wooded land as having either a crown cover of 5–10 per cent of trees able to reach 5 metres at maturity, or a crown cover of more than 10 per cent of (dwarf or stunted) trees not able to reach 5 metres.

FOUNDATION

Foresters shared the Enlightenment's great hope in rationality. They envisioned a sylvan future in which rational use of the varied forests could be continued for ever. They called their vision the principle of sustained yield.

Survey, assessment, measurement, experiment and analysis became their modes of conduct. Wood was the forests' major product and they needed to know how much they had and how the various ways that they could tend the forest – silviculture – could alter the amount they could grow. Then they needed to select the best way, given that the consequences would occur far into the future. They found an economic model for this, although it was of dubious assistance. Most importantly, they developed models to regulate the output from each forest as a whole. The following four chapters present what can be called the foundations of modern, scientific forestry.

The foundations were laid from the end of the eighteenth century in the German kingdoms and principalities, France, the Austrian empire and the Swiss cantons. They addressed the problem of how to restore the extensive native forests by managing them, rather than by creating plantations as on British estates. The problem was critical because industrialisation in most of Central Europe had to be fuelled by wood, rather than by coal as was increasingly the case in Britain.

The problem was at first local because ground transport was too expensive to cart wood far and beyond that wood could only be taken by river or sea, or converted into charcoal that could be carried further. The increasing demand for wood could be met in only two ways: by extending the frontier of exploitation ever further into the mountains with the aid of ingenious forest engineering; or by searching for rational ways to increase continual production from the accessible forests.

Rational ways needed rational governance formalised in specific forest legislation. Although some early endeavours can be traced to the sixteenth century, legislation was increasingly strengthened and standardised across Europe in the nineteenth century. This was particularly important for the regions mining for salt and iron ore in Austria, Hungary and Slovenia where demand for fuel wood and charcoal was intense, but it extended to some Scandinavian and Mediterranean countries. By the end of the century, fifteen European countries had modern forest laws and had established forestry organisations to implement them.[1]

The search for rational ways to manage the forests started in the larger German states of Saxony, Prussia, Bavaria, Württemberg and Baden, but applied in the smaller ones. Large areas of the forests were in their princes' hands and provided the main part of their states' finances. Both forests and finances had been drained by the Seven Years War (1756–1753), the Napoleonic Wars (1799–1815) and other conflicts. Repair was taken in hand in the German-speaking world by a new form of public administration, 'cameral science'. Its students were trained in economics, mathematics and law in universities and staffed the state bureaucracies. Their quantitative regime collected descriptive statistics to assess the natural resources and regulate their states' economies.

It was from this cameralist regime that forest science and a new type of forester emerged.[2] Practical and hunting knowledge, the domain of the traditional foresters and huntsmen, was still required but, for the new regime, mathematics, surveying, forest measurement, hydraulic engineering, botany, zoology and tree physiology, soil science and other subjects had to be covered.[3] Modern foresters had to be applied scientists.

Several small forestry schools were started in the second half of the eighteenth century, but it was not until the nineteenth century that major state-funded institutions were established. The first was within the Austrian empire in 1807, at Banská Štiavnica in Slovakia. Two of the most notable were the Forest Academy at Tharandt in Saxony founded by Johann Heinrich Cotta, formally established in 1816; and the Prussian Forest Academy established at the University of Berlin in 1821 and moved to the forest area of Neustadt-Eberswalde in 1830 under the direction of Friedrich Wilhelm Leopold Pfeil. France followed suit by establishing a national forestry school at Nancy in 1824. Eight years later the first scientific forestry journal, *Allgemeine Forst- und Jagzeitung* [General forestry and hunting magazine] started to be published in Frankfurt: clearly forestry was standing as a distinct scientific discipline. By the end of the nineteenth century, eighteen European countries had forestry schools – several in Germany and two in Britain – and the United States had just started its first two.

The leaders of the forestry schools, such as Cotta, Pfeil and Georg Ludwig Hartig made the early advances in quantitative methods but, as forest science expanded, special experiment stations needed to be set up to concentrate on research. By the 1890s there were eight of these in Germany and six in other European countries. They dealt with common species and problems and needed an international forum, not only to exchange scientific information, but also to standardise research methods.

Wednesday, 17 August 1892 marked forest science's formal entry on the international stage as the German, Austrian and Swiss research stations formed the International Union of Forest Experiment Stations (now the International Union of Forest Research Organisations, IUFRO).[4] The foundations on which modern forestry would stretch across the world had been laid.

Notes

1. The first of what can be considered modern forest laws was passed in what is now the Czech Republic in 1754. This was followed by Denmark 1805, Switzerland 1809 and Germany 1816.
2. Lowood 1990.
3. The traditional huntsmen, the *Jägers*, occupied an important place in German forest culture. Brown 1887 details the curricula in several forestry schools.
4. International Union of Forest Research Organisations (IUFRO) 1992.

1.

MEASURING

Scientific forestry could not start until the foresters knew how much wood they had and how fast their trees were growing. They had to measure them. It seems such a basic step, and they had such prosaic means at first that it hardly appears to qualify as 'science' when compared with the chemistry and physics of the time; yet its very simplicity highlights a strand that runs through the history of forest science: the means to understand the complexity of the forest are always more limited than human hopes.

As in the other sciences, the foresters used models to devise ways to estimate what they needed to know from their measurements. There were stages in this from single trees, to forests, to growth. Over time, better instruments, aerial photography, computers and satellite imagery enabled the foresters to improve their models from simple graphs to intricate computer systems. Measurement was an essential subject in a forester's training. Known as 'forest mensuration' it became the subject for research and a steady production of text books into the twenty-first century.[1] The foresters' measurements and models of wood, the subject of this chapter, were not extended significantly to the other uses and values of the forest until late in the twentieth century, as described in later chapters.

Four basic measurements

Forests are large, awkward things to measure. They are endlessly complex and some are vast. It is hard to get around in them; most are full of hills, creeks, steep slopes and rocky outcrops; it is difficult to see far and in places the undergrowth can be so dense that a path has to be cut. Many things needed to be measured to gain a quantitative understanding, but the foresters started by concentrating on the wood. They took only four basic measurements: area, and the diameter, height and age of the trees.

To measure even the simplest attribute, the area of the forest, needed a considerable amount of work. First, surveyors or sometimes the foresters themselves had to survey and map the boundaries. This could define individual forests until the progress of national mapping systems during the nineteenth and twentieth centuries could provide an overall 'cadastral' map of ownership and use. The boundaries of each forest had to be marked on the ground and this 'demarcation' was essential to assert legal ownership. The foresters had to re-assert it endlessly against farmers edging their boundaries into the forest, or loggers thieving the trees. It is an issue that filled the records of mediaeval courts and endures today, severely so in some tropical forests. It highlights another strand that runs through the history of the forest science: it needs steady governance before it can be applied.

When the external boundaries had been clearly established, the foresters had to map the interior to find the area covered by the different types of forest and by the non-forested parts – clearings, roads, rivers, lakes, rocky outcrops and so forth. In the mountainous or hilly country of most forests, detail of the ridges and streams was also needed. This all had to be done on foot. However, it became easier and more accurate when aerial photographs started to become widely available in the 1950s. The men who slogged through the world's forests in the nineteenth century could never have imagined the men and women sitting at distant computer screens examining satellite images of forests they hardly visit in the twenty-first.

The main trees in a forest are so large that the foresters could only take three basic measurements – diameter, height and age – and there are so many trees in a forest that they could never measure even these three on all of them; they could only *sample* them. They had to decide how many of which measurements to take on which trees, but the number they could take was limited by the time available. Moreover, their sampling methods had to be easy enough for their assistants to apply. Sampling systematically such as measuring every tenth or hundredth tree selected on a grid through the forest was simplest. As the theories of analytical statistics were developed from the 1920s, more sophisticated sampling methods became available that involved selecting the samples at random. They were more efficient, but they needed more highly trained assistants for both measurement and calculation.

In an exception to sampling, some Swiss, Slovenian and French foresters measured the diameter of *all* the trees at five or ten-year intervals. They started doing this in Switzerland in 1890 and in some forest estates in Slovenia two years later as part of a new silvicultural and management system, the 'Control Method' discussed in later chapters.[2] In the 1960s and 1970s they adopted sampling methods as part of general measurement reforms.

Diameter was easiest to measure because the foresters or their assistants could reach their arms around the trunks with a tape to measure their girth, or use a calliper to measure their diameter at 'breast height'.[3]

It was their first basic measurement, but it was liable to all the vagaries of branches, bumps and buttresses that had to be allowed for, and in the tropics they had to allow for large buttresses that could stretch high up the trunks of some trees (Figure 7.2).

Height was their second measurement, but they could only take the time to measure a smaller sample, because measuring height took much longer than diameter. Either a simple staff had to be used, as shown in Figure 1.1, or the angles and distance away had to be measured so that the height could be calculated geometrically, with due allowance for sloping ground. In dense forests it was hard to find a clear view of the tops of the trees, or of their bases when there was undergrowth.

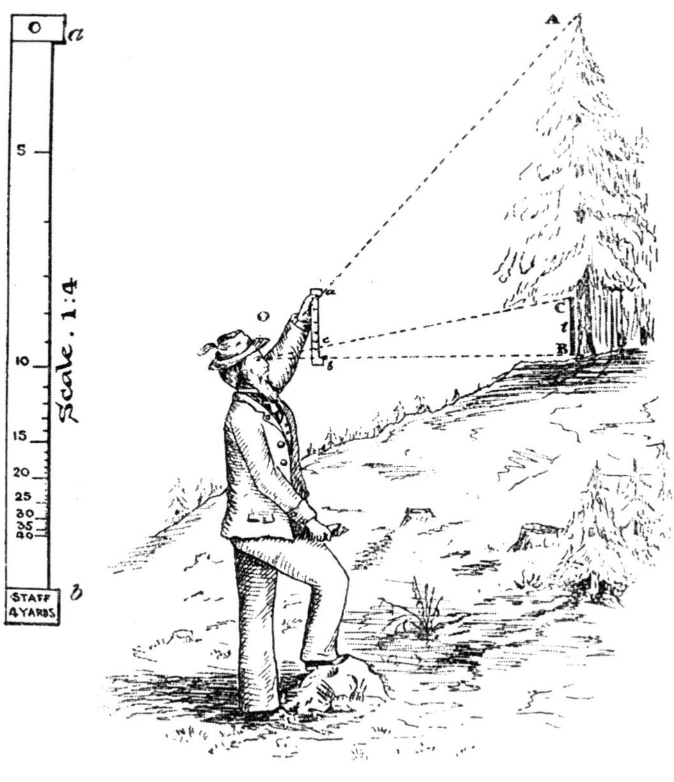

Figure 1.1. Using a simple staff, Christen's hypsometer, to measure the height of a tree.

Source: W. Schlich 1895. *A Manual of Forestry, Vol III Forest Management including Mensuration and Valuation.*

Age was their third measurement. In plantations they knew age from records of when the trees had been planted. In natural forests they could count the distinct annual rings on the stumps of felled trees or on thin cores bored out of standing trees. Counting the rings worked well in most of Europe and North America, and for some deciduous species like teak in those parts of the tropics with dry seasons, but not in many other regions of the world where the seasons were less marked and the trees did not form distinct annual rings.

Trees

The foresters needed to know the volume of wood in a tree standing in the forest with only its diameter and height measurements to go by. For this, they had to make tables for everyday use that they could look up to find the volume. Their first step was to fell a sample of trees of different sizes. At least fifty or so were needed to prevent a few oddly-shaped ones distorting the overall pattern. Once the trees were on the ground, they could remove the bark and measure the diameters at regular points along the stems. They assumed that different parts of the stems were geometric shapes and applied the standard formulas to calculate the volume of each section and thus find the total. Real trees, of course, varied from such models and some pernickety foresters cut a few up and submerged each piece to measure its volume by the amount of water displaced. They even built special tanks, 'xylometers', to do this, but they found that the formula method was good enough, provided that the diameters were measured frequently enough along the stem, commonly every thirty to fifty centimetres.

In the very essence of scientific inquiry, the foresters sought a general model of tree volume based only on diameter and height. They easily calculated the cross-sectional area or 'basal area' of each tree at breast height from its girth or diameter. When they combined this with the height, they found that they could draw a single line to represent the volume of all the trees in their sample. From this they could draw up the tables. As calculating methods and machines improved they could replace their hand-drawn graph lines with equations.

They had to repeat the whole laborious process for each of the major species. They characterised the shape of each species as it varied from region to region by a 'form factor'. By measuring the diameter at a point well up the stem on a sample of trees, they could find the factor and reduce the amount of work needed when they came to assess the stands.

Stands

In their endless struggle to understand the forest's complexity, the foresters introduced two categories to describe its variation: stands and site index. A formative category was the idea that a forest was made up of discrete 'stands': patches that had the same species, or mix of species and which were approximately the same age and size for practical purposes. Stands could be delineated on maps and foresters could manage similar stands in the same way.

The usefulness of stands as a descriptive category varied considerably from forest to forest. In some forests, the divisions between stands were readily apparent, due to natural changes in the topography or soil. They could also be due to the pattern of past storms or fires or, especially in Europe, to the patterns of past human use. In other forests, the divisions were not distinct, as patterns blended into each other without clear boundaries, or patterns might not be recognisable in any useful manner. However, the orderly practices of scientific forestry as they were increasingly applied, served to create a human pattern of distinct stands in many forests.

Site index was their second category. It rested on the idea that stands of the same dominant height at a set age could be sorted into classes. It proved a useful concept for classifying the similar forests according to what the foresters were most interested in: their productivity. Apart from trees grown in the open which tended to be short, they found that they could gauge the potential productivity of a stand from the height of its tallest trees at a set age, commonly twenty or twenty-five years for plantations and fifty or a hundred years for natural forests. They drew graphs to show the heights sample trees reached at different ages and arranged them into usually five to seven classes. These enabled the foresters to place a stand of any age into its appropriate category. The classification worked best in plantations or in natural forests where most trees were of the same age and species.

Forests

Determining the total amount of wood standing in a whole forest was always the most politically pressing problem. Experienced foresters and timber men were used to assessing the amounts by eye but once they could find volume by measuring standing trees, they could set about making their assessments more accurate. Except in the forests mentioned earlier, it was not practicable to measure all the trees and they had to devise sampling methods using strips or plots.

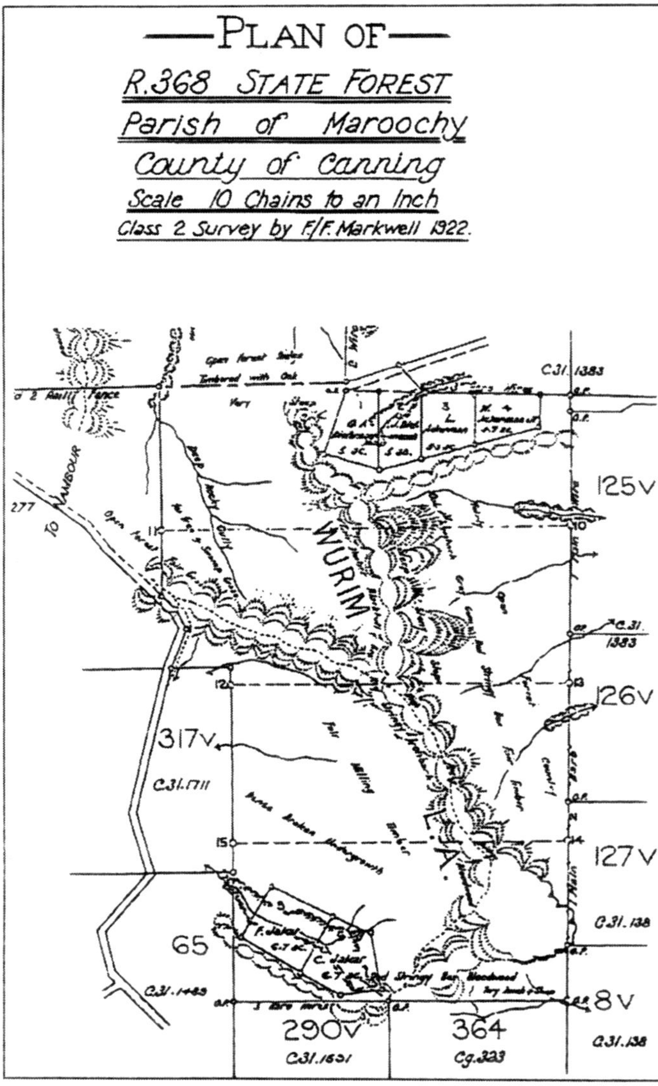

Figure 1.2. Map of part of Maroochy State Forest in Queensland, Australia. Strip survey by Forest Foreman Markwell in 1922. Original scale of 10 chains to 1 inch (1:7920). Numbered portions are private properties. Dashed lines show the strip survey lines, 20 chains (400 m) apart.

Source: Queensland, Department of Primary Industry–Forestry.

Systematic strip surveys were the simplest way of sampling a forest and were the initial means of assessing them in countries where little was known beforehand. Crews of three or four men crossed the forest on lines 200 or 400 metres apart and recorded the species and diameter of every tree within twenty metres either side of the centre line. They also sketched in the position of the ridges, creeks and tracks and noted other features of the landscape and forest on the maps. These simple surveys provided crude estimates of the volume of wood standing in the forest and were sufficient for initial exploitation, but not for scientific management.

As the foresters subsequently made more intensive surveys, they were able to delineate the different stands and to judge or measure their site quality. They could then make their assessments more efficient and show the resources of each type of stand or quality of site. Rather than strips, plots were selected through the forest and carefully laid out, square or round in shape and usually between 0.04 and 0.1 hectares in size. The diameter of every tree and the height of some of them were measured so that the foresters could calculate the volume of wood per hectare on each plot and come to more accurate estimates of the total amount.

Walter Bitterlich, an Austrian forester, made a breakthrough in the process of measuring every tree to find the basal area of a plot in the 1940s when he developed the 'angle-count' method. For this, the foresters only needed a small glass prism or an instrument called a 'relascope' to take a sample of a stand's basal area from where they stood. It was a great saving in time and effort. The relascope was developed into a sophisticated instrument that could also be used to measure the height and diameter of points along the stem of a standing tree. This was particularly useful where a diameter well up the stem was needed to determine the form factor.

Measuring trees and plots became far easier as instruments using laser technology became available in the twenty-first century and as electronic devices that recorded the measurements digitally replaced written records that had to be transcribed later. Moreover, locating and re-locating plots with global positioning systems (GPS) was easier and more accurate than the simple surveys they replaced.

In spite of all the effort it took to assess total amount of wood in the forests, it was not enough. The foresters had to know how fast the trees were growing if they were ever to make a scientific base for the sylvan future they envisioned.

Box 1.1. Walter Bitterlich (1908–2008)

Walter Bitterlich was descended from several generations of Austrian foresters. He went to school in Innsbruck and Salzburg and studied forestry at the Hochschule (now University) für Bodenkultur (Applied Life Sciences), BOKU in Vienna. He became a forester in the Austrian federal forest service (ÖFB) from 1935 to 1941 and then served in the German army in Russia and Normandy. He resumed his career with the forest service in 1949, where he remained until 1966. During this period he published his method of angle-count or point sampling for basal area (*Winkelzählprobe*), a method that revolutionised forest mensuration. His first ideas about it are documented in his diary as early as 1931.

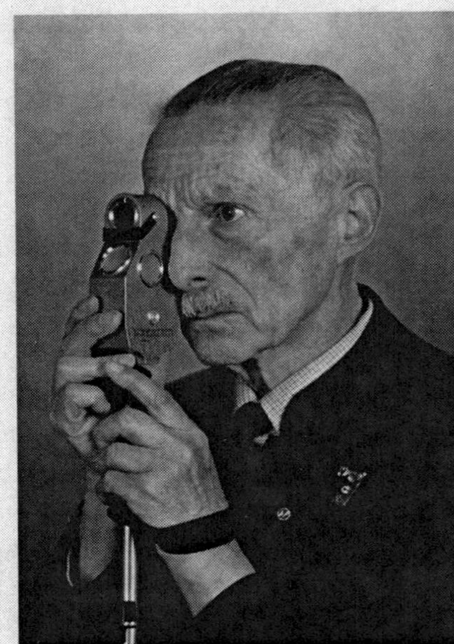

Dr Walter Bitterlich demonstrating a Spiegel-Relaskop.

Photograph courtesy of Relaskop-Technik, Saltzburg.

From 1950 he worked closely with the instrument manufacturers, FOB, now Relaskop-Technik, in Salzburg on the industrial development of his patents. This resulted in the production of the Spiegel-Relaskop and Tele-Relaskop instruments. These enabled the basal area of a stand to be measured quickly and the diameter of trees to be measured at any point up their bole without having to climb or fell them. They became the state-of-the-art instruments for forest inventory worldwide.

In 1966 Dr. Bitterlich was appointed as a professor at the University für Bodenkultur, BOKU in Vienna. After his retirement in 1978, he continued his scientific work, published widely, including *The Relascope Idea* in 1984, and continued to develop new instruments, including the multiple use Bitterlich Sector Fork. He received awards in Austria, from forestry associations all over the world and was made an honorary member of the Society of American Foresters. For recreation, he enjoyed painting.

Growth

Measuring the rate of growth was much more difficult than measuring the standing volume, particularly as the foresters wanted to know how the rate might change over the life of the trees, a hundred or more years into the future. There were three things that they could do, but none of them was entirely satisfactory.

The first was to search a whole region for any areas where the ages were known and measure their height and volume. This information could be plotted on a graph and used to sketch lines of how the forest might have grown. In natural forests, the foresters might be able to find the age of a stand from counting the annual rings on a felled tree but, where annual rings were not formed, they could only rely on memories or records of when a particular stand had once been felled and regenerated. When they wanted to develop plantations of introduced species in a new region, they could usually find places where they had been tried on a small scale.

The second method was to cut cross-sections at points along the stem of a few felled trees and count the number of annual rings at each point. The changing diameter, height and shape of each tree could be drawn. Although their volume at past ages could be calculated, the method was of most use in conjunction with the first method for showing how height had grown with age. It could only be used in forests where annual rings were developed.

The first two methods inferred, rather than measured, a forest's rate of growth. To measure it directly, the foresters had to turn to their third method of measuring the same stands at long intervals, commonly every five years. They set out permanent sample plots – sometimes called 'growth' or 'increment' plots – commonly about 0.05 hectares in size, in which they measured every tree. It was a long-term enterprise but, as they could not wait a hundred years to collect all the measurements on particular plots, they started with plots at different ages. After a few years of measurements, they could draw lines to show the growth pattern of each plot and then combine them to show the pattern of growth on each type of site. This provided them with more soundly based estimates than they had been able to prepare by their first method. Tables showing the standing volume of each type of site at various ages were prepared from the graphs for convenient use.

Quantitative base

By the end of the nineteenth century, it was clear that the quantitative base for forest science, with an accepted set of measurement processes and protocols, had been firmly established. In the twentieth century, instruments and techniques were steadily improved, sampling was refined, statistical analysis replaced graphs and the sheer mass of measurements being gathered was processed effectively.

The advances in sampling and statistical analysis made in agricultural science from the 1920s and 1930s were taken up by forest scientists.[4] They prepared their volume and yield tables using equations calculated by 'regression', a form of statistical analysis. Their laborious work was eased as mechanical calculators became available that could multiply squares as well as add them and even calculate square roots.

Statistics carried the great hope that the variety of nature could not just be tabulated, as the cameralists had done, but measured. Individual trees might vary widely, but collectively they created patterns, or 'distributions' – 'normal', skewed', 'exponential' – that could be described mathematically. Statistics brought the unsettling realisation that nothing about the forest could be stated with certainty, although it carried the reassurance that the degree of confidence in what was stated could be given.[5]

The advent in the 1960s of the now ubiquitous electronic computers and their increasing power enabled the foresters to expand how they took, processed and analysed forest measurements. They could devise new systems for assessing the forests, such as the Continuous Forest Inventory System with which they monitored the condition of large forest regions by installing a network of plots, each about 0.1 hectares, measuring their trees and recording their health every five to ten years. It had a precedent in the labour-intensive European Control System, mentioned earlier. However, the impetus came not from Europe but from North America where the foresters could accumulate all their measurements in large computer 'data bases', and examine them more intensively than had ever been possible from the paper records. No longer did they have to use some of the abstractions, like site index class to simplify their calculations, they could take a statistical view of the entire forest.

The confident development of forest science in the USA was marked in 1955 when the Society of American Foresters launched *Forest Science* as a new quarterly journal of research and technical progress, distinct from the *Journal of Forestry* that it had published since 1902. *Forest Science* set rigorous standards for the statistical analyses in the papers that it accepted. Although it reviewed German language publications on volume tables in its first year, it included very few non-English language publications in later years. This reflected a general and increasing shift that made English the dominant language of science. Whereas in the nineteenth century many English-speaking foresters had studied in Germany or France and had acquired the languages to do so, textbooks in English, such as Schlich's five-volume *Manual of Forestry*, were available from the start of the twentieth century.[6] In consequence, the confident developments of forest science in the second half of the twentieth century became partially separated by language groups.

The great hope that science would provide an enlightened understanding of the forests was founded on the measurement and statistical analysis introduced in this chapter. The textbooks do not display doubt: they instruct. The journal articles present their analyses, replete for the last fifty years with calculations of the confidence that can be placed in them. The forest, however, is not so easily reducible and the simplifying assumptions that had to be made in order to proceed with measurement were just that, 'simplifying'. Four in particular had significant consequences for how forestry was founded.[7]

First, and most obvious, was the foresters' prime focus on trees and the quantity of wood they could supply for construction and energy, the critical economic development question of the times. Although they had always dealt with the forests' other attributes in practice, they only gradually extended their science in the twentieth century to ecological attributes.

Second, but far less obvious now, was the notion that the forest was composed of stands of trees, each stand or patch being more or less homogenous in terms of its trees. How far this was a reasonable assumption depended on the type of forest and its history. It obviously applied for small plantations where all the trees were the same age and planted at the same time but, in many forests with several species of trees of different ages, the distinctions and boundaries between one stand and its neighbours were less clear or had little meaning.

Third was the way in which volume tables and later equations were prepared from the measurements of a sample of the trees of a particular species. These worked well for conifers and young trees with more or less regular shapes, but did *not* work well with many irregularly shaped hardwoods and old trees.

Fourth was the most fundamental: the forest that the foresters measured was not the forest that they were growing. From the start of systematic forestry, the young stands that they were producing were generally more uniform than those they replaced. This was most pronounced in the twenty-first century when plantations were established with seedlings produced as clones by intensive breeding techniques (Chapter 8). Nothing like them had been grown before.

Measuring and analysis were defining attributes of science but, although great advances were made in founding forest science, the forest itself could never be fully known.

Notes

1. In English-language text books, Schlich 1895 included mensuration with management in volume 3 of his *Manual of Forestry* and Jerram 1939, 1949 provided a text when Schlich's became out of print. Texts published in the USA followed Schlich and Jerram: Graves 1906; Chapman 1921 with subsequent revisions with Demerritt 1931, 1936 and with Meyer 1949; Bruce and Schumacher 1930, 1935. The expansion of forestry in developing countries from the 1950s led to agencies such as FAO issuing introductory handbooks. Publication continued as mensuration developed in the twenty-first century. Husch, Beers and Kershaw 2003 is a recent example and Pretzsch 2009 provides a comprehensive treatment of the mensuration and modelling needed for intensive silvicultural practice.
2. See Chapter 2 for the silviculture of selection forests and Chapter 4 for their regulation by the Control Method.
3. 1.3 metres or 4 feet 3 inches above the ground in most countries, 4 feet 6 inches in the USA.
4. The advances were led by Ronald Fisher's work at the Rothamsted Experiment Station in the UK and marked by the publication of his textbooks on *Statistical Methods* in 1925 and *The Design of Experiments* in 1935, both volumes with many subsequent editions.
5. In an early example of calculating volume equations on a computer, the volume of a natural forest of klinki pine on a forest concession at Bulolo in Papua New Guinea was estimated in 1958 to be known within +/- 15 per cent with 95 per cent confidence, although it was hoped that further analysis would improve the estimate. Dargavel 2008, pp. 186–7.
6. Schlich 1889–1925.
7. A substantial review of the assumptions from the perspective of hindsight is given by Puttmann, Coates and Messier 2009.

2.

TENDING

The modern foresters' great task, their very *raison d'être*, was to renew the forests after the trees had been felled, but first they had to restore those misused for grazing or cut wantonly for timber. Their vision of a sylvan future depended on such restoration and on the endless cycles of renewal that they would manage. Behind their vision lay a deep sense of the forest's power to heal itself: a storm topples a great tree, herbs and grasses appear in its place and in a year or two new trees show themselves. It was part of a forester's daily life in all its infinite variety. From the 1920s ecologists developed theories of plant succession to account for it and from the 1970s conservation biologists included it in their wider concept of ecosystem resilience. In the eighteenth and nineteenth centuries the foresters devised ways of tending the forests, 'silviculture', to renew them and to produce just such shapes and sizes as were needed. The two core issues of silviculture – regeneration and spacing – were advanced by the science described in this chapter.

The foresters tended the forests in their care in other ways, as they had from ancient times: here they built a fence to keep young trees safe from grazing animals; there they burned piles of dry branches before summer; everywhere they kept roads and tracks free from fallen branches and cleaned their ditches; then they collected seeds and strewed them in the forest, or sowed them so that they had young trees ready to plant two or three years later. All these and many more were the mundane tasks of every day, interrupted when wildfires or storms had to be dealt with and aided as the years passed as much by experience and technology as by formal science.

The problems of how to tend the forests were not new and the foresters founded their modern science by changing or developing past practices. Four of the old European methods can be mentioned here: coppicing, the practice of picking and choosing trees, the French *tire et aire* method and the clear-cutting method.[1]

Villagers all over Europe had woods of hardwood trees such as oak, sycamore, hazel and chestnut which they cut close the ground every few years, allowing a new crop of coppice shoots to grow from the stumps. They grew the shoots until they were big enough to provide the poles, fuel wood or other products they needed. Seven to twenty years were commonly needed before the stumps, or 'stools' were cut again. The process could be repeated and some stools are reported to be centuries old.[2] In many places the villagers were required to keep a number of trees, the 'standards' to grow on through several cycles of cutting the coppice so that they could provide large timber eventually.[3] When they no longer depended on the coppice woods for their fuel, fences and huts, the foresters had to find ways to convert them to modern uses.

An equally old but enduring practice posed a greater challenge. In France it was called *jardinage* or gardening and in America 'high-grading' to describe the simple process of picking over the forests to choose the trees required at the time. It could be lightly and carefully done, but driven by the market and practised for a long time it degraded the forests by leaving only the poorest types of tree to grow. For example, many of the Australian eucalypt forests contained a mixture of species, sizes and qualities of trees. Without a significant market for fuel wood in many areas, only the trees containing good sawmilling logs were taken. As the remaining trees quickly spread their crowns and roots, they left few chances for young trees to establish successfully.

The French *tire et aire* method, with origins in the fourteenth century, became a national requirement under Colbert's seventeenth century reforms. It specified the number of trees 'of lofty growth, of great vigour and fine proportions' that were to be reserved from felling and whose seed would enable the forest to regenerate naturally afterwards.[4] Importantly, it envisaged that the foresters would select the areas to be felled, or 'coupes', in an orderly way that moved across the forest, beginning with those that were already partly regenerated.[5] The foresters would return to each coupe twenty years later to clear away any older trees interfering with the new crop.

In practice the *tire et aire* method left too many senile trees in the forest and much of its regeneration came from coppice shoots, rather than from the new seedlings needed for the best quality trees. A hundred years after Colbert's reforms, Henri-Louis Duhamel du Monceau examined the problems. He was ideally placed to do so as he was the French navy's Inspector-General, overseeing its supply of timber; he was a member of the Academy of Sciences; he had a long-standing interest in trees and between 1755 and 1767 published an eight-volume text on wood and forests.[6] Duhamel saw that drastic measures were needed to restore the forests: all the trees should be cut, 'clear-cut', the ground should be cleared by burning all the undergrowth, a crop of lucerne should be grown to prevent weeds and enrich the soil and a new crop of trees planted. This clear-cutting method did not work well for the oak forests he was most interested in because the young oak seedlings needed some shade to get properly established.

The foresters evolved different types of silviculture from such past practices and adapted them to the complexities of their particular forests. For example, in alpine areas they had to guard against avalanches; on steep slopes and vulnerable soils they had to prevent erosion; and they had to be alert to the danger of wind blowing trees down when they opened the forest canopy by felling. But what they could do in practice was always governed by their markets and funds: unless they could sell the wood – or promise to do so in the future – they could not afford the silvicultural work. Large trees of the best species, with a good straight form and free from timber defects, could always be sold, but less favoured trees were more difficult or impossible to sell. However, conditions were favourable for developing silvicultural systems to a high standard during the late eighteenth and nineteenth centuries in parts of Europe due to the high demand for wood. The foresters could usually have most of their trees felled and sold when and where they wanted. When coal, oil and electricity gradually replaced wood as the main source of energy in the twentieth century – or earlier where there was abundant coal as in Britain – these circumstances no longer applied; nor did they apply in the New World while its over-abundance of timber lasted; nor in the tropical forests where few of their many species could enter the world trade.[7]

Research

Silviculture in its modern sense was founded in the nineteenth century on the base of the earlier methods. What distinguishes it is the use of quantitative measures to guide its development. We can see the transition in the life of Karl Gayer whose background in mathematics was important to his appointment as a forester in the 1840s when the cameralist outlook was firmly established and the scientific approach was becoming the foundation for forestry.

Nineteenth century research varied from measurements of simple forest trials, to designed experiments and laboratory work on particular aspects. Many regional and individualistic methods had been devised by the separate forest administrations of the countries, kingdoms and major principalities of the German-speaking world. They were described and named in the forestry literature of the eighteenth and nineteenth centuries

Box 2.1 Karl Gayer (1822–1907).

Johann Christian Karl Gayer.

Picture from *Allgemeine Forstzeitschrift* 1957, (10): 137; Wikimedia Commons.

Johann Christian Karl Gayer was born in Speyer on the banks of the Rhine in 1822. He was the son of the county archivist, Peter Otto Bernhard Franz Gayer who died when he was only twelve years old. Karl Gayer attended the High school (Gymnasium) in Speyer and started to study architecture and mathematics at the Polytechnic school in Munich. Because of financial problems he was not able to finish his studies. In 1843 he started to work as a forest assistant in the Bobenthal and Bienwald forests. In 1845 he was transferred to the public forest administration in Speyer and in 1851 was promoted as the forester in charge of the Weisenheim am Berg district. He was clearly highly competent because four years later he was appointed both as the county forester based in Speyer and as the Professor of Forest Science at the Royal Bavarian National Forest Academy, then based in Aschaffenburg. As well as fulfilling the duties of these positions, he carried on his research in the Black Forest and in the forests of Spessart and Odenwald.

In 1878 parts of the Forest Academy were moved from Aschaffenburg to become part of the University of Munich. Gayer was appointed as the Professor of Forest Production and was awarded the degree of Doctor *honoris causa* of the Unversity. He became Rector (i.e. President) of the University from 1889 to 1890. Gayer's most important books, *Die Forstbenutzung* [Forest Utilisation] (1868) and *Der Waldbau* [Silviculture] (1880) were re-published several times and remained standard textbooks for a long time. When Sir William Schlich prepared his five-volume *Manual of Forestry* he used a translation of *Die Forstbenutzung* as an English language textbook for forest utilisation in 1896.[21]

He developed his silvicultural principles for selection forests in a further book in 1886 and set out economic goals in another in 1884. His thesis for how the forests should be managed was adopted for the Bavarian State Forests in 1890.

After retiring, he continued to be a member of the forest faculty and published several books. His work was recognised with several awards and decorations.

and became, with the French methods, a necessary part of forestry education. For science to advance and be applied elsewhere the methods needed to be classified, but their complicated and overlapping terminology, expressed in the compound nouns of the German language, made it difficult for foresters elsewhere to understand.[8] It was not until 1928, when R.S. Troup published his comprehensive *Silvicultural Systems*, that the English-speaking world had an accessible classification and text book.[9] Troup drew his examples from Germany, Switzerland and France, with a few from England, Canada and India. The classification provided a 'language' for the concepts used.

Forest researchers found that moving from the trial and error methods by which silviculture had evolved to a quantitative, experimental basis was difficult and expensive. For example, they needed areas that were reasonably uniform, large enough to have different treatments and free from other operations for many years. In some countries, designated experimental forests were attached to forestry schools and state experiment stations. Research activity increased and extended into more countries as more forestry schools and research institutions were set up. By the 1930s the International Union of Forest Research Organisations (IUFRO) had members in 38 countries, of which only eight (Canada, India, Indonesia, Japan, Mexico, The Philippines, South Africa and USA) were outside Europe. Germany with its well-established university forestry schools and research stations was pre-eminent but research in the USA was becoming increasingly important, with a string of federal research stations set up across the country from 1908. Japan started a research organisation in 1905. In British India the establishment of the Forest Research Institute at Dehradun in 1906 was followed by the appointment of silvicultural research officers in several provinces. Research there, in Indonesia, The Philippines and some African colonies started on the difficult silvicultural problems of tropical forests.

Some specialist foresters, who became known as 'silviculturalists', conducted numerous investigations in different types of forest. These provided a theoretical basis to the existing practices, from which some authors deduced general principles. For example, the American silviculturist, Frederick Storrs Baker revised his authoritative 1934 *Theory and Practice of Silviculture* as *Principles of Silviculture* in 1950. By the mid-twentieth century, it was clear that silviculture had moved from description to principle.[10] However, it was also clear that the idealised systems of the text books had to be adapted to individual forest circumstances.

The systems for tending the forests fell into two great camps – 'even-aged' and 'selection' – depending on often contested ideas of what was desired and what was possible. In even-aged systems all the trees in a stand were felled at the same time, or over a short period, to create a new crop of trees of roughly the same age. In selection systems, individual trees were picked out for felling and their companions left to grow for another ten or twenty years when the process would be repeated. This left each part of the forest always full of trees of different ages, and usually of different species, growing together.

Even-aged systems

The more uniform the wood, the better it suited industrial production. Plantations of the same species, or 'monocultures', that produced crops of the same age, quality and size were the industrial ideal. The scientific advances that underpinned their development – from the spruce plantations of Central Europe, to the pine plantations of France and Scotland and those that will supply one-third or more of the world's wood in future – are discussed in later chapters. Here we are concerned with how even-aged silvicultural systems were developed for the natural forests. This served not only to make them more valuable to industry, but also to make them more manageable; they were simplified.

Even-aged systems were usually only possible in natural forests where most of the trees could be sold. This was not only an affair of markets, but varied greatly with the type of forest. Conifers, with their straight stems, were often easier to sell than the generally less regular hardwoods and the great coniferous forests that spanned the northern world were typically dominated by only one or two species. Moreover, as a result of past fires or storms they were commonly found naturally in even-aged stands. A few hardwood forests, such as the mountain ash forests of Australia, the world's tallest hardwoods, were also even-aged as a result of catastrophic fires.

Seed, the survival of tender new trees and wind were the foresters' major concerns. Would there be enough seed from the trees being felled to grow a new crop? It was a perennial question and one that engaged silviculturists everywhere. For example, researchers in Finland investigated how often the trees produced seed in different parts of the country, while in Sweden they investigated how much it varied according to the age of the mother trees.[11] When there was not enough seed every year, the foresters could collect some in bountiful years and sow it in lean years or raise it as seedlings to be

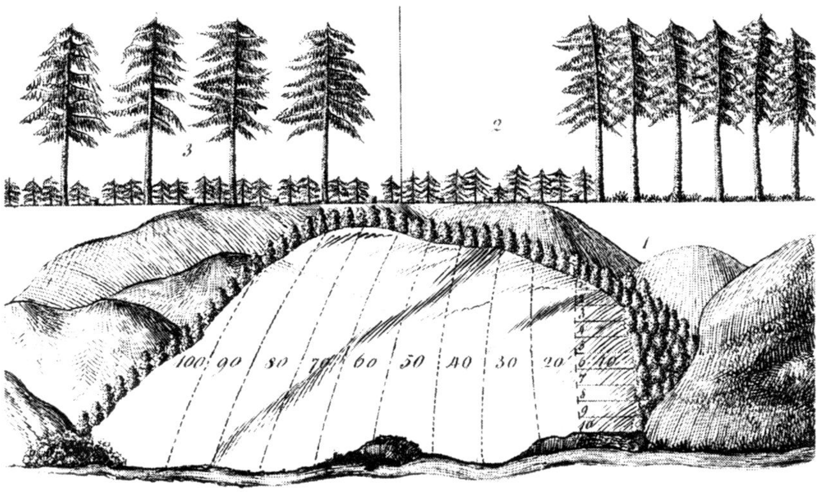

Figure 2.1. Two silvicultural systems.

1. The lower section of the picture shows how a forest could be laid out systematically working on a hundred year cycle. It is divided into ten parts each to be cut in a decade. The first part on the right hand side shows how it is divided into annual coupes.

2. The upper right section shows a clear-cutting system in which young trees are growing after all the mature trees have been removed. Adjacent to them on the right are mature trees.

3. The upper left section shows a shelter-wood system in which young trees are growing in the shelter of a few mature trees after most of the mature trees have been removed. The retained trees would be removed after the young trees were firmly established; note how they have extended their branches.

Source: K. Kasthofer 1828. *Der Lehrer im Walde* [The Teacher in the Forest].

planted out later. Their other strategy was to keep a few of the old trees as 'seed trees' to shed their seed later. Silviculturalists researched how far the seeds of different species dispersed from the seed trees, then calculated how many seed trees they needed to keep in each hectare and investigated whether cultivating or burning the forest floor to prepare 'the seed bed' would help more take root. They conducted laboratory and glasshouse studies of the best ways to store seeds and to germinate them ready for sowing.

Although 'clear-cutting' to remove the old trees at one go, even leaving some scattered seed trees, was attractively simple – as in the *tire et aire* method – it exposed the young seedlings to the full forces of the climate. The importance of this varied widely between species. Several systems of sheltering them while providing more seed were devised. One was to fell the forest in alternate strips, fifty metres or so wide. Small blocks in a chequerboard pattern could be used instead. The uncut strips or blocks could be removed when the young trees were tall enough to provide some shelter to their neighbours.

Another way was to fell the old trees in stages spread over several years, enabling the young trees to become established under a 'shelter-wood'. This might be done over a whole area in what became known as the 'uniform system', or in scattered gaps, or in such a gradual way that the young crop had some variations in its age.[12]

Selection systems

Selection systems that kept a mixture of ages and species in every part of the forest were all that were possible when markets were too limited to do much else, or when the most desirable species grew best amid other species.[13] But in other forests they could be chosen. Although even-aged forests suited industry and were the simplest to manage, they were not necessarily the most desirable. A strand of 'nature-based', or 'close to nature' forestry arose in Central Europe towards the end of the nineteenth century whose foresters argued that silviculture should be based on ecology rather than industry.[14] It arose largely in reaction to the clear-cutting forests of mixed species and ages and the widespread practice of replacing them with even-aged spruce plantations.[15] For example, the Professor of Forestry at Munich, Karl Gayer wrote in 1884 that: 'We have lost the path to nature ... we must toil along the return journey to the selection forest'.[16] The vision of restoring the forests and managing them in perpetuity remained, but the means and priorities were changed. The change affected measurement, economics and regulation as well as silviculture and, by the twenty-first century, was to become widely accepted in Europe, at least in principle.

Selection systems could be applied most successfully in forests where the young trees could tolerate being partially shaded by the older ones. Foresters working in limestone, or 'karst' regions of Europe were the most notable exponents as the forests there were composed of the shade tolerant silver fir and beech trees which grew mixed with other trees on land that

Figure 2.2. The selection system in the Regensberg communal forest, Switzerland, managed to maintain a mixture of spruce, fir, beech and other trees in a mixture of ages.

Photograph: John Dargavel.

was particularly liable to erosion. To avoid erosion and wind damage they needed to keep the land continuously covered by forest and could only remove scattered single or very small groups of trees at any one time. They faced two difficulties. Practically, it was difficult to fell and remove a large tree without damaging the others; and the scattered gaps had to be large enough to allow a group of young trees to grow. Much depended on the skill of the foresters and the fallers. The foresters had to mark which trees were to be removed so as to maintain each stand with the ideal structure of species, ages and sizes. Theirs was an almost intuitive skill based on experience. The fallers had to be equally skilled in order to fell large trees without breaking the younger trees around them.[17] The science to quantify selection systems and regulate their yield was developed later (Chapter 3) but personal judgement was always needed in practice.

Spacing

When the foresters controlled how closely the trees grew together – a stand's 'density' – they could influence the trees' form: if they kept a stand dense, the trees would be straight and thin and most species would shed their lower branches; if they opened it up by removing some, the remaining trees would grow to larger diameters, but with more branches and bigger knots in the wood. If they removed too many, the site would not be fully occupied, total production would be reduced and the remaining trees might blow over.

In plantations, they could control spacing from the start by choosing how far apart to plant the young trees, but in natural forests their over-riding concern was to ensure that they had enough of them, although sometimes they were over-abundant, coming up like a 'wheatfield'. Until the young trees in either plantations or natural forests had grown enough for their crowns to touch one another – to make a 'closed canopy' – the foresters had to ensure that the young trees were not crowded out by weeds. After that, spacing could be controlled by thinning out the trees in a stand.

The foresters could sell the trees removed – thinning was rarely possible otherwise – which provided some welcome income long before the main crop was ready. They had to choose how many trees to remove, the 'intensity of thinning', and how many times to thin during the course of a rotation. They also had to choose the type of tree to be removed: the smallest, weakest ones or the largest ones in greatest demand. They could cut whole rows, sometimes every third or fifth, when they first thinned a plantation.

Forest science entered a new stage in its development when it started experiments to measure the effects of thinning on the growth of forests and the quality of the trees. With so many options, the design of thinning and plantation spacing experiments was difficult. However, the German and Austrian researchers were able to prepare a 'Guide for thinning and incremental felling trials' that was adopted by IUFRO in 1903 to assist researchers in other countries.[18] The experiments started to be designed more rigorously from the 1930s as statistical designs were adopted. The UK took a prominent place in this as researchers from there, France and Germany produced a guide for forest growth and yield investigations that IUFRO adopted in 1936.[19]

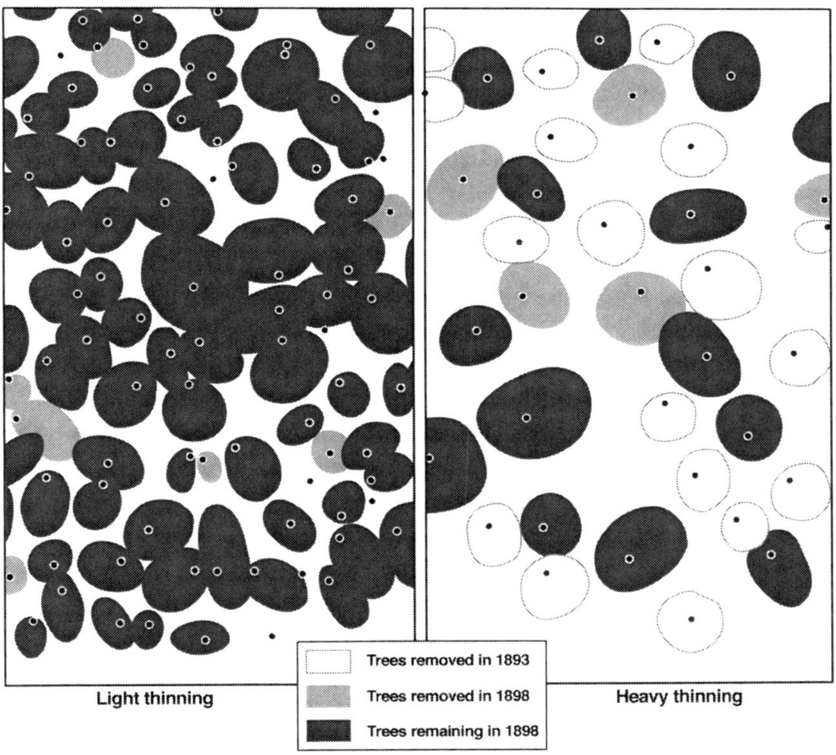

Figure 2.3. Crown maps of an Austrian thinning experiment in a stand of beech, conducted in 1893 and 1898. Redrawn from Böhmerle 1900.[20]

Key: Trees removed in 1893 - light area; in 1898 - cross hatched areas; remaining in 1898 - dark areas.

Even with the guides, forest experiments were notably difficult. The researchers needed to find sites that were as uniform as possible and replicate each treatment with several plots to allow for irreducible variation. Some experiments were repeated in different regions to see if the responses to treatments changed. It required a great deal of effort to establish each experiment and more to maintain and re-measure all the plots, typically for ten or twenty years, or longer when possible. Moreover, the measurements needed to be preserved and later retrieved. The path from design to analysis was hazardous: plots might be lost to storms or wayward loggers, original markings might be lost and re-measurements uncertain, research fashions or institutions might change, records might be lost or be difficult to interpret years after they were made and so forth. As only completed experiments were ever reported; we will never know how many fell by the wayside.

The availability of computers from the 1960s and the advances in what was becoming known as biostatistics enabled some specialist forest researchers not only to analyse the measurements in more sophisticated ways, but also to link separate analyses into 'growth and yield models'. These composite packages enabled foresters to gauge the growth of each stand and the amount of wood it might yield, just as the previous graphs and yield tables had, but also to gauge the possible effects of different spacing and thinning options. Some models included the possible effects of improving the genetic base or of increasing growth with fertilisers (Chapters 8 and 9). The sophistication of the computer modelling that had been achieved by the 1980s was concentrated on the intensively managed industrial plantations, such as those in the Southern United States, Australia, New Zealand and parts of Europe. These large, even-aged monocultures, mostly of conifers, but with increasing areas of eucalypts and a few other hardwoods, were the least variable of all forests. Much less progress was made in natural forests with their mixture of ages and species and least progress was made in the tropics.

The foresters' hopes that silvicultural research would provide them with a quantitative basis for deciding how to tend the forests in their charge were realised most successfully in European and some North American forests and in the industrial plantations. For these forests, they could make reasonable estimates of the amount of wood or sizes of trees that they might grow by selecting different provenances or species, by spacing the trees differently or by thinning the stands to different intensities. With estimates of the likely effects of the possible options before them, they had to decide which to select; and that was a matter, some thought, for economics.

Notes

1. Bown 1883, Troup 1928, Reed 1954.
2. Rackham 1996, Figure III, gives an example of an ash stool at least 800 years old.
3. For example the French Forest Ordinance of 1669 (discussed in Chapter 1) specified that 16 trees per arpent (47 per hectare) were to be kept as standards on each hectare, Brown 1883. In Tudor England legislation specified 10 per acre (27 per hectare).
4. A minimum of 10 trees per arpent (30 per hectare) were to be reserved. Brown 1883.
5. Troup, 1928, p. 127.
6. Bonnaire 2000. Duhamel published extensively on several topics as well as his forest treatise, *Traité complet des Bois et des Fôrets* in 1755–1767. His insights into the difference between phenotype and genotype in volume 5, *Semis et Plantations des Arbres et de leur Culture* 1760, are discussed in Chapter 8.
7. Sieferle 2001.
8. The complications of terminology are discussed by Spurr 1956.
9. Troup 1928. R.S. Troup was Director of the Imperial Forestry Institute at Oxford University.
10. Baker 1934, 1950.
11. Ilvessalo 1927, Petrini 1928.
12. Troup 1928, pp. 31–2 records multiple German and some French terms being used for the uniform system in different areas and periods.
13. African mahogany and Australian red cedar are examples of non-gregarious species.
14. Johann 2006.
15. Johann *et al.* 2004.
16. Karl Gayer in 4th edition of his textbook on silviculture, cited by Köstler 1956, p. 211.
17. Köstler 1956, pp. 215–6.
18. International Union of Forest Research Organisations 1992, p. 5.
19. The Guide was published as *Outlines for Permanent Sample Plot Investigations*, cited in International Union of Forest Research Organisations, 1992, p. 9.
20. From the experiment conducted by Austrian Forest Research Institute, Mariabrunn, 'Interrupting the leaf canopy of a beech stand', Sample plot 7, Laabach; Böhmerle 1900. Böhmerle's thinning experiments with spruce, fir, black pine and beech that started in 1882 were the basis for further research related to growth and yield. He was one of the founders of IUFRO in 1892.
21. The volume was prepared by W.R. Fisher (1896) using a translation of the 4th edition of Gayer's *Forstbenutzung* and incorporating other material from France, Britain and India.

3.

PROFITING

The Enlightenment carried the great hope that the complexities of the natural and social worlds could not only be understood through science, but also managed rationally through the emerging insights of sociology and economics. However, the foresters found that deciding how their vision for the future could be managed was no easy task. How they looked to the powerful but limited insights of economics for help is the subject of this chapter.

The ancient question of how long to grow trees before they should be cut down was the first to be addressed. Everybody knew it took a long time to grow big timber and country people knew how long it took to grow the small sizes they used for fuel, huts and fences. No doubt estate owners like John Evelyn and his brother were just as aware of prices as they were of sizes and quantities but their decisions were made from experience or *ad hoc* to meet their needs at the time. They had no economic theory to guide them and arguably needed none.

Foresters in the nineteenth century had to make their decisions in quite different circumstances. The cameralists' insistence on quantitative assessments of the state's economy and resources was driven in part by the need for better methods to value properties for taxation – an ancient state problem. The subsequent development of forest measurement and experimental methods had provided European foresters with familiar and increasingly reliable *physical* measures – area, diameter, height and volume of wood – on which to base their decisions. Although they could quantify some of the benefits like timber or grazing that were specific to particular users, others, like water, were widely diffused and some, like beauty, lay outside the realm of quantitative science. Moreover, all were spread over the long cycle of the forest's life, which made valuing particularly difficult.

The changing scale of the demands for wood altered the economic situation. Railways and roads were being built and water transport systems were being improved, enabling wood to be traded further away. These changes were experienced unevenly but wood, rather than being primarily a local commodity, became a regional, national and an international commodity for some forests. Although the amount of wood needed for industrialisation and urbanisation continued to increase, the type of wood demanded from the forests changed. The market for hardwoods fell as charcoal was replaced by coal in making steel and the market for fuel wood reduced. By contrast, the market for softwoods increased markedly as building and construction work expanded. Producing wood became an important business, so important that other uses were relegated to the status of 'minor products'.

Politically, the doctrines of *laissez faire* called for less state control and some states in Central Europe modified their previous forest regulations in favour of industrialisation policies. This, combined with various crises in state finances, resulted in large areas of state-owned forests being sold to investors, mines or industrial companies. Austria, for example avoided state bankruptcy by selling its forests.

These and other influences were behind a change in the forest landscape across much of Europe as large areas of the broadleaved forests were replaced by planting conifers, largely Norway spruce and to a lesser extent Scots pine. The new stands, planted as even-aged monocultures, replaced forest stands that were naturally mostly uneven-aged with a mixture of species.[1] The increase in the proportion of conifers was also boosted by converting moors and heath lands to conifer plantations. These changes and their consequences are discussed in later chapters; here we are concerned with the extent to which economic analysis informed the decisions that foresters made.

The mentality of the new industrial capitalism centred on the *profits* that could be obtained from *investment.* It started to dominate the mentality of forest owners who needed an economically rational basis – a calculus – for their forest and investment decisions. However, the foresters' mentality was grounded in regenerating the forests and making them as productive as they could. Their quantitative understanding was greatest for the even-aged systems and plantations and it was for these that their economic analysis was developed.

Once they could measure the wood production and display it graphically, they could show that the rate at which young stands grew each year, called the 'current annual increment' increased until a stand reached a certain age after which it gradually declined. However, the long-term production, expressed as the *average* amount of wood grown to each age, called the 'mean annual increment' peaked at a later age after which it declined. In some forests in Saxony, for example, spruce reached its maximum annual increment at seventy years of age but did not reach its maximum mean increment until

Figure 3.1. Method of sliding logs into rivers used in Austria and parts of Germany to exploit mountain forests of conifers.

Source: W.R. Fisher's revised 1896 edition of Karl Gayer 1868 *Forstbenutzung.*

twenty years later; in the slower-growing oak forests, the relative ages were one hundred and thirty, and one hundred and forty years.[2]

The foresters had to decide what was the best age, or 'rotation', to fell a stand of trees and replace it with young ones. Should it be when it was growing fastest? Or should it be a few years later when it had reached the maximum rate over its lifetime? Or should they wait for many, many years until it had carried its absolute maximum volume and decay had set in? Their vision of a forest future was not only that the yield should be sustained in perpetuity, but also that it should do so at the highest level possible: they should choose the rotation that had the maximum mean annual increment.

Foresters like Friedrich Wilhelm Pfeil and Johann Christian Hundeshagen became attuned to the new economic mentality. They thought that the rotation should not be decided on such physical measures alone; the costs and returns should be taken into account so that the best value, or 'land-rent', could be obtained. Rather than find the rotation age that was most *physically* productive, they wanted to find the age that would be most *profitable* for land owners and investors. This was difficult to work out because

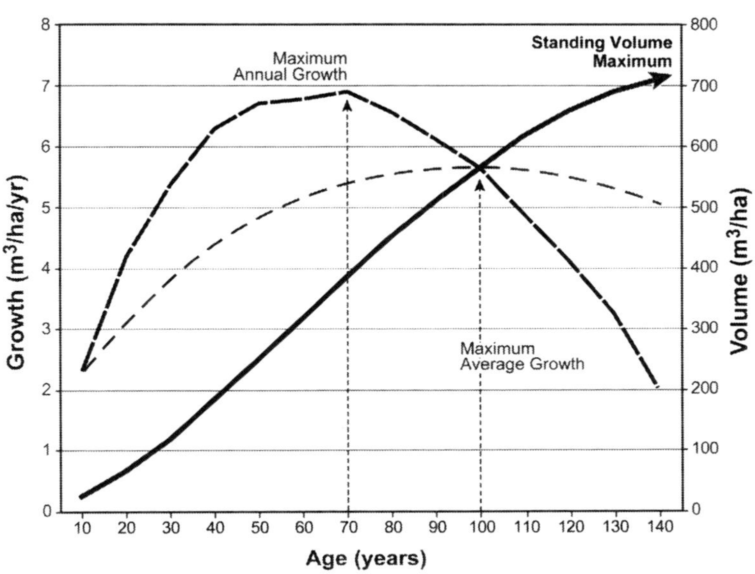

Figure 3.2. Annual, average and total growth of spruce.
Plotted from W. Schlich 1895. *A Manual of Forestry, Vol III Forest Management including Mensuration and Valuation.*

the cost of regenerating or re-planting a stand, and the cost of maintaining it every year, would not be recouped until the trees were grown and their wood was sold decades later. It was tricky because wood was produced from thinning during the rotation as well as from the major sale at the end; and there was a major difficulty because the foresters envisioned a succession of crops into the distant future. It all made for a triple problem: a mathematical one of making the calculations; an economic one of defining 'profit'; and a practical one because foresters had to decide about their existing forests as well as the series of crops that would follow. Several methods were advocated and strenuously argued in lengthy and idiosyncratic papers and books that often confounded the three aspects of the overall problem.

It was not until 1849 that the German forester, Martin Faustmann was able to bring his knowledge of economics to bear on the rotation problem.[3] He had two key insights. First was that the process used by bankers could enable the several costs and returns that occurred at different times in the forest to be turned into commensurate economic values – ones that could be added together – by 'moving' them backwards or forwards through time in an analysis. Ever since the seventeenth century, bankers had known how to value long-term loans by calculating the 'compound interest' that is incurred when the interest due each year is added to the debt and incurs further debt itself. They also knew how to calculate the value of annuities. Their techniques enabled Faustmann to compound the planting cost, and the intermediate returns from the thinning sales, to the end of the rotation where their values could be added to the return from the final felling, before their total was discounted to the present.

Faustmann's second and mathematically elegant insight was that the succession of identical crops that the foresters envisaged could be seen algebraically as a series stretching to infinity, thus allowing an exact value to be calculated simply – although generations of forestry students have rarely appreciated it.[4] His insights contributed to the development in economics of what is now called 'discounted cash flow analysis'. The value that he and others called 'land rent' – also 'land' or 'soil' 'expectation value' – is now included in the general economic term 'net present value'.

Faustmann's insights were refined by others including Max Robert Pressler who in 1860 proposed that the 'yearly annual percentage' return on the capital invested would suit investors better than the land-rent which was the return on the land.[5] He knew or assumed the value of the bare land and then used the same processes of compounding and discounting to search

Box 3.1. Max Robert Pressler (1815–1886)

Max Pressler was born in central Dresden in 1815 where he attended high school and technical college, graduating with an engineering degree. In 1836 he obtained a position as senior teacher in a vocational school at Zittau in Saxony, close to the borders of both Poland and what is now the Czech Republic. After four years there, he was appointed as a professor of agricultural and forestry engineering, and mathematics in the Royal Saxon Academy of Forestry that had been established in 1811.

Max Robert Pressler

Photograph from *175 Jahre forstliche Ausbildung in Tharandt* (Dresden, 1986), p. 26, Wikimedia Commons.

In 1858–1859 he published his first book, *Der rationelle Waldwirth und sein Waldbau des höchsten Ertrags* [The Rational Worth of Forests and the Silviculture of the Highest Income], and in 1860 set out his ideas in a long journal paper in *Allgemeine Forst- und Jagd-Zeitung* that enabled them to become widely known. His economic ideas were welcomed in Saxony and some other states as supporting the establishment of spruce monocultures to supply the industrial demand for wood.

Like Faustman and other forest scientists of the period who developed new measuring instruments, Pressler developed a drill – the Pressler Increment Borer – that is still in widespread use today. It enables a researcher to extract a thin core from a tree so that its age and growth rate can be found by counting and measuring its annual rings without having to cut it down. He also developed simple instruments and practical ways of measuring timber and cattle that farmers could use.

Pressler was made one of Saxony's Privy Councillors and was awarded an honorary doctorate by the University of Giessen.

for the interest rate that would balance the commensurate costs and returns. He had no elegant solution and the calculations had to be repeated several times until the analyst could home in on the correct rate – now known as the 'rate of return'.

Interest rates

The foresters' greatest problem in using economic analysis was, and remains, the exponential effect of compound interest over the long time they need to grow trees. The critical question was what rate of interest they should use. The economists' logic was that it should be the rate at which an individual investor could borrow or lend the money. Market rates of five or more per cent were common, although the foresters often made the calculations, rather wishfully, at rates of three per cent or lower.

The effect is displayed in a hypothetical example in Figure 3.3. If an investor borrowed $1,000 to invest in forestry, it would incur a debt that built up increasingly steeply until it could be repaid by selling the wood. If the interest rate was three per cent, the debt would be $4,400 by the time the trees were fifty years old, but if it was seven per cent it would be ten times as

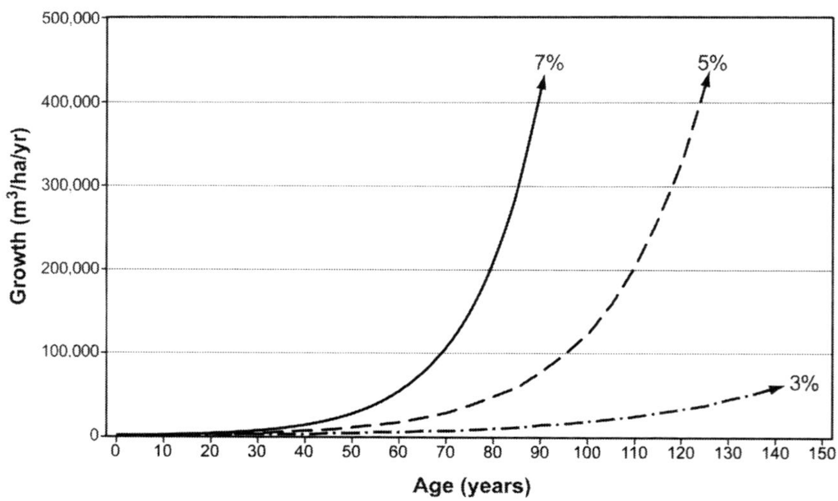

Figure 3.3. The burden of the debt incurred at various ages by an investment of $1000 at compound interest rates of 3%, 5% and 7%.

much. Foresters often grow trees for longer. In Austria, they are not allowed to clear-cut stands before they are sixty years old and they generally grow them for one hundred years. In France or Germany they grow oak for 150 years that in this example would carry a burden of debt of $84,000 at three per cent or an impossible $25 million at seven per cent.[6] Growing trees for such a long time was clearly a problematic business at commercial rates of interest, but what might be profitable? Could the foresters grow trees faster than debt? If so what was the optimum age to fell them?

A few foresters laboriously calculated net present value, or the rate of return and the mean annual increment for every possible rotation, using the yield tables, costs and wood prices relevant to their forests. Because of the uncertainty about which interest rate to use, they calculated the values using different rates and showed that the economic value peaked at a much earlier age than did the mean annual increment. The most profitable rotation was not the most physically productive, and vice versa. As they tried different interest rates, it became clear that the higher the rate, the shorter the rotation needed to be. Alarmingly, they found that their calculations gave negative results for the longer rotations and higher interest rates; growing trees on very long rotations would never be profitable.

Decisions

Economic calculations were one thing, making the decisions was quite another. Although the differences between the criteria were clear, German foresters argued about which to choose.[7] Shorter, more profitable rotations meant that there would be less wood produced each year than if the foresters continued with their aim of growing stands to the age of their maximum mean increment. Not only was the economic argument at odds with how the foresters had envisioned their ideal future, it required them to extend the basis on which they made their decisions beyond the physical measurements they were used to. Not surprisingly, its acceptance was ambiguous, ranging from ubiquitous in the textbooks to least in practice, and varying between private and public forests.

Landowners and foresters in much of Central Europe had already realised that growing the hardy and adaptable Norway spruce in pure stands met the immediate demands for wood quicker and was a better business than growing the slower-growing hardwood species like oak, or the naturally mixed stands of conifers and hardwoods, as mentioned earlier.[8] The Faustman/Pressler economic models were developed *after* the conversion to secondary spruce forests had begun. They reflected and measured the economic effects of the conversion decisions being made and doubtless guided others.

In the private forests, individual estate owners had many objectives that did not fit neatly into the economists' assumption of simple profit maximisation. The most eminent forester in the English-speaking world, Sir William Schlich, recognised this when he declared in 1895 that:

> It rests with the owner to determine in each case what the objects of management shall be, and it then becomes the duty of the forester to see that these objects are realised to the fullest extent possible and in the most economic manner.
>
> In some cases the realisation of indirect effects [beauty, climate, erosion, game, hygienic] requires a special and distinct management, but in the majority of cases they can be produced in combination with economic working.[9]

The case was rather different for the owners of woodlands and small forests who used their holdings to provide work in the non-agricultural seasons, or for a reserve against hard times or contingencies, or for other individual purposes. Their decisions were not necessarily driven by demand or profit. Collectively they were particularly important in countries where a high proportion of the national forest area was held in small individual forests.

In the public forests, the decisions were less able to be fitted into the Faustmann model than in the private forests. At least in the private forests there was a single decision-maker, the owner, whereas in the public forests the ostensible single decision-maker, the state, was liable to be swayed by those that had material interests or who advocated ideas about how the forests should be conserved and managed. Until the rise of the environmental movement in the 1970s and 1980s, the foresters' vision was the most prominent idea, and one that was well entrenched in state forest services, research institutes and professional societies. The extent to which its influence prevailed against those who wanted to clear the forest for agriculture or simply exploit its resources varied across the world, probably being greatest in Central Europe.

Economic theory, as the eminent American economist, Paul Samuelson, noted, does not debar the options of devastating the forests by not regenerating them, or of converting them to some other use if they are more profitable than growing more trees.[10] The conflict with the foresters' vision of perpetual forest was expressed in debates about the discount rate. The long rotations – fifty to one-hundred years or more in many forests – needed to grow large, good quality logs could only be justified in economic theory at very low or even negative discount rates. Most private owners knew it without any model, as the British forest manager, W.E. Hiley summarised in 1930:

> Despite the fact that some of the most intensive forest management in the world is to be found on private estates, it may be accepted as a generalisation that private or commercial ownership of forests, when unfettered by legislative restriction, generally leads to devastation.[11]

The foresters' often-confusing arguments fell into two main groups. One sought different ways of accounting for forestry costs and revenues, the other sought justifications for using very low discount rates. Foresters in the first group were often responsible for the practical day-to-day management of large estates. Their problem was one of commercial viability. They had to find a way of generating a cash flow each year from the forest while steadily building up its productive capacity. For example, Eugen Ostwald developed his relative forest rent theory while managing Riga's large city forests.[12] Rather than consider each stand individually, he considered the forest as a whole and took the growing stock as its major capital value. The path to profitability of the enterprise was to be found by choosing the stands to be felled that contributed most to the profit of the enterprise each year. Hiley, with a sharp eye to the cash flow, stressed the control of costs and developing markets for products that could be obtained early in the rotation.[13]

Foresters in the second group raised all sorts of arguments to escape from the 'power of compound interest' over public forests. To Samuelson's amusement, one of the U.S. Forest Service Chiefs asserted in 1925 that no discounting was necessary provided that revenues exceeded costs in a steady-state or normal forest.[14] More general were arguments that an especially low rate should be applied to allow for wider objectives than short-term wood production.

Limitations

The foresters' hope that economics could provide the discipline for making their silvicultural decisions in a rational way was limited by the assumptions it made about the complexity of the human world, and by the complexity of the forests.

Conventional or 'neo-classical' economists posited a rational order to human affairs and proffered their vast, complex and evolving literature with an ever-more ingenious calculus.[15] They assumed for their starting point that society was composed of well-informed, greedy and impatient individuals unswayed by ethics, loyalty or religion, whose autonomous action in a 'free' market maximises the total welfare of society. They saw the rate used for discounting as the way in which an investor balanced greed – wanting to have more rather than less – with impatience – wanting to have it sooner rather than later. They came to mould these assumptions in a host of ways to allow for the cost of information, risk, consequences to third parties or 'externalities' and the 'imperfections' of real markets where governments regulate or corporations exert their power.

Faustmann and Pressler accepted these simplifying assumptions and made some specifically for the rotation problem. They assumed that: there was a 'perfect' capital market in which money could be lent or borrowed; all the prices, costs and interest rates were constant and known with certainty; forest land was a commodity that could be bought, sold or rented without restriction; there were no restrictions on cutting; and that decisions could be made about each stand without considering the forest as a whole.[16] In spite of these limitations, the Faustmann/Pressler model became a standard analytical tool for *comparing* silvicultural options such as rotation length, thinning regimes or fertiliser applications.

There were greater limitations that came from the inherent complexity of the forests, and greater still from the forests' interaction with the atmosphere. There were wider objectives than producing wood and maximising profit. How these eventually influenced the course of forest science and the economics of forestry in the environmental era and in the new millennium is discussed in later sections of this book. During the foundation phase, the greatest limitation of the Faustmann/Pressler model was its restriction to analysing decisions about stands, when for larger and public forests there were overriding concerns about managing whole forests, the subject of the next chapter.

Notes

1. Johann *et al.* 2004 discusses the history of these complex changes in detail.
2. Schlich 1895, vol.III, pp. 386, 389.
3. Martin Faustmann (1822–1876) was a German forester and forest scientist. He started to study Catholic theology at the University of Giessen in 1841 but soon shifted to forest science. He finished his studies in 1848. From 1846 on he contributed papers to the forestry journals *Allgemeine Forst- und Jagd-Zeitung* and *Jahrbücher der Forstkunde*. He and his wife developed an instrument named the *Spiegelhypsometer* to measure the height of trees. He became the leading forest officer of Dudenhofen near Darmstadt and lived in nearby Babenhausen untill he died. His most famous paper, 'Berechnung des Wertes welchen Waldboden sowie noch nicht haubare Holzbestände für die Waldwirtschaft besitzen', *Allgemeine Forst- und Jagd-Zeitung* **15** (1849) was translated into

English as 'Calculation of the Value which Forest Land and Immature Stands Possess for Forestry', first published in M. Gane, ed. (trans. W. Linnard) 1968. *Martin Faustman and the Evolution of Discounted Cash Flow: Two Articles from the Original German of 1849.* (Oxford: Commonwealth Forestry Institute) and republished in *Journal of Forest Economics* in 1995.

4. The Faustmann Model (Land-rent theory) is:

$$\text{Present Value} = \frac{\text{Value of timber at rotation age} - \text{Cost of planting } (1 + \text{discount rate})^{\text{rotation age}}}{(1 + \text{discount rate})^{\text{rotation age}} - 1}$$

The value of timber from intermediate thinning operations is compounded from when they occur to the rotation age and included with the value of the final felling. The capitalised cost of annual maintenance is calculated as = annual maintenance cost/ discount rate and is deducted from the Present Value.

5. Pressler [1860] 1995.
6. Some intermediate returns could be received from thinning out some of the trees during the course of the rotation, but they were generally minor.
7. Möhring 2001.
8. Johann *et al.* 2004.
9. Schlich 1895, p. 1.
10. Samuelson 1995, first published in *Economic Inquiry* **14** (1976), reprinted in *Journal of Forest Economics* in 1995.
11. Hiley 1930, p. 227.
12. Markus 1967.
13. See Hiley 1930 and especially his 1956 book on plantations.
14. Samuelson 1995.
15. Marxist economics construes the world in terms of class conflict and the accumulation of the surplus value of labour, but to the best of our knowledge has not developed analytical techniques for the forest problems discussed in this chapter.
16. Johannson and Löfgren 1985, p. 75.

4.

REGULATING

Control was essential: the forests could not sustain their yield in the long-term, nor could markets be supplied steadily, unless the foresters could control the amount of wood being cut. They had to be able to limit the supply in order to ensure it in the long-term. Inevitably, they had to fight industries looking only for short-term profits. Science, and the authority and confidence that it gave them, was their major armament. Increasingly, the foresters knew what they were talking about.

It was a political, cultural and scientific problem. The political impetus for forest control was not radical in Europe, given its long history of legislation, and the foresters held a well-established, militaristic place in Central European culture, but scientifically it posed another thorny problem. Knowing how each stand grew was not enough for control; there were just too many different species, qualities and ages in most forests and, although the foresters could work out the best time to fell each stand, the yields of wood rarely fell due when needed. Just as they had to manage each stand as a whole, as well as its individual trees, they had to manage each forest as whole, as well as its individual stands. The idea that they could regulate cutting was nothing new, as coppice woods were already managed on regular rotations, but it was the introduction of quantitative control that laid one of the foundation stones of modern forestry.

What each 'forest' was that they had to manage was never problematic in the settled landscapes of Europe where the tenures were known and individual forests had names. A forest might be only 142 hectares like the Swiss communal forest of Regensberg, or it might be 11,000 hectares like the French state forest of Tronçais, or it might be 678,000 hectares like the Willamette national forest in the United States, but it was the unit of ownership whose future was to be planned, its yield sustained. It was in this sense autonomous, if only in the hopes of foresters.

The foresters developed various forms of quantitative control to regulate the yield of wood from the forests depending on the type of forests and on how extensive their measurement information. This was best for the even-aged forests, but more difficult for the uneven-aged forests.

Even-aged forests

For the earliest and simplest system of control, the foresters needed to know only the *area* of the forest. If they divided the area into the number of equal parts corresponding to the length of the desired rotation – say one hundred parts for a hundred-year rotation – and they allowed only one part to be cut each year, then they could establish a regular pattern.[1] Evelyn had heard of something like it being practised in France, he thought on an eighty-year rotation, in the seventeenth century.[2] How it might be laid out in the forest is depicted idealistically in the lower section of Figure 2.1.

The weakness of the system was that supply varied from year to year because parts of the forest yielded different amounts of wood depending on their quality and condition. It could be overcome if the foresters assessed the forest and knew the standing *volume* of the different stands. Then they could divide the forest into one hundred parcels of roughly equal volume, cutting larger areas for the poorer qualities and smaller areas of the better ones.[3] Five- or ten-year periods could be used instead of single years, to accommodate variations in demand as well as supply in both these systems.

Both systems were adapted for the conversion of poor, patchy forests that had been heavily cut or grazed in the past and which did not have enough healthy trees for growing wood. The foresters could make some more or less arbitrary decision on how long to plan a shorter 'conversion period' in which to replace the existing forest with the more productive one. Clear-cutting the existing stands in such an orderly manner and establishing a new crop met the need to increase long-term wood production.

The Normal Forest ideal

A German forester, Johann Christian Hundeshagen, formalised the concept of the 'normal forest' in a book published in 1826.[4] It proved to be an influential ideal in the foundation of modern forestry, and one that lasted through most of the twentieth century. It can perhaps be claimed to underlie what in the twenty-first century we call sustainability. In nineteenth century forestry, 'normal' carried its scientific meaning of an abstract ideal, or standard rather than its common meaning of 'usual'.

The idea derived from the area regulation system and foresters explain it most easily that way. For example, if a forester had 1,000 hectares of a perfectly uniform forest in which ten hectares were felled and replanted every year in the same way for a hundred years, and the trees were only cut down

BOX 4.1. Johann Christian Hundeshagen (1783–1834)

Johann Christian Hundeshagen

From *Biographien bedeutender hessischer Forstleute* (Wiesbaden und Frankfurt am Main 1990), p.341, Wikimedia Commons.

Johann Christian Hundeshagen's career reflected the tensions of the Enlightenment era when both science and more liberal policies were replacing the old aristocratic order. He was born in Hanau in the Electorate (i.e. principality) of Hessen, in what is now Germany, towards the end of the eighteenth century. After graduating from high school he undertook a two-year course at the state's new forestry school at Waldau and Dillenburg where the famous forester Georg Ludwig Hartig was teaching. He followed this with cameral and natural science studies in the University of Heidleberg. With these qualifications he was appointed as a state forester, first in the forest and salt mine office at Allenburg and then in Hersfeld, where he remained for ten years.

His advancement within the state administration was blocked because aristocratic connections counted for more than forestry training and experience, a situation he resented as unjust. He took up a position as a professor in the University of Tübingen where for the next sixteen years he produced several major works on forestry. His *Methodologie und Grundriss der Forstwirtschaft* [Methodology and Main Features of Forestry] (1919) and *Encyclopädie der Forstwissenschaft* [Encyclopaedia of Forest Science] (1821–31) which summarised the entire contemporary knowledge of the German Classical School of forestry and thus had a tremendous influence on the development of forestry in Germany and abroad.

However Hundeshagen's work was not without controversy. An applied science field like forestry that was closely associated with hunting did not sit easily in this ancient university, famous for its philosophers, theologians and astronomers. Faculty differences were as nothing compared to the political storms of the period. In 1819 he joined his fellow professors and students in a political action to free some farmers by storming a building where they had been jailed.

Leaving Tübingen, he moved first to Fulda and then to the smaller University of Giessen.[8] As well as being a professor responsible for teaching students, he was also in charge of the forest institute. However, relations with the state administration were so difficult that he resigned and devoted himself to teaching and writing his major work that set out the normal forest ideal *Die Forstabschätzung auf neuen wissenschaftlichen Grundlagen* [Forest Estimation Based on New Scientific Findings] (1826). The years of work and controversy took their toll and he died in Giessen in 1834 at the age of fifty.

and replanted when they were one hundred years old, a normal forest would have been created that would provide a yield that could be sustained for ever. Inherent in this simple ideal is the perfect balance between the amount that the resilient forest grows each year, its increment and the amount being cut. Moreover, the total amount of wood standing in the forest, its total 'growing stock', made up of the amount in each age class, would always stay the same.

No real forest was ever like the normal forest ideal; forests were too variable and their growing stock was usually below, but sometimes above the ideal quantity. The *amount* that could be cut was the major industrial and often political interest. Controlling it, rather than the area, was the foresters' essential quantitative tool for ensuring the forest's future, raising its productivity and providing steady supplies in the meantime. With their normal forest concept – of balancing growth to the amount being cut – in mind, they had to devise ways to calculate what they called the 'allowable cut' from whatever information they had about the standing volume and increment.[5] No mathematically exact solution was possible and much depended on the amount of information available.

If the total standing volume of wood was all the foresters knew, they could use a simple method devised by Wilhelm Wolfgang von Mantel in Bavaria. All they had to do was to divide half the standing volume by the length of the rotation to make a crude estimate the annual 'allowable cut'. The method assumed that the forests were 'normal' and dangerously overestimated the amount that could be safely cut in forests whose increment was in fact low due to their poor condition.

If the foresters had already mapped the forest into site quality classes and had made yield tables, they could use the forest's increment as well as its growing stock to make their estimates. They knew how much wood they had in the real forest's growing stock and they knew what the growing stock of a normal forest ought to be. They had to devise ways of allowing for the increment that would occur during the period they thought it would take to convert their forest into an ideal state. Several formulas were advocated such as the Austrian, Hundeshagen and Heyer methods that we can mention here.

The Austrian method set the allowable cut based on the increment that a normal forest should have, but reduced it where the real forest had less growing stock than a normal forest should have – the common situation in Europe where the forests had to be built up to restore their productivity. The allowable cut could also be increased – a common situation in the New World, but one that often needed to be explained. Their old stands carried high volumes, but their increment had declined. Younger stands would be more productive in the long-term, but it was difficult to strike a balance between converting the plentiful yields from the old stands to the productive potential of their replacements.

Figure 4.1. Clear cut blocks in the Douglas fir region of Western Washington and Oregon, USA, c. 1950s-1960s.

Source: Weyerhauser Co. photo Courtesy of the Forest History Society, Durham, NC, USA. FHS1232.

Hundeshagen developed a similar way of setting the allowable cut, but based it on the ratio of the increment of the real forest to that of the ideal one. It had to be recalculated every few years. Carl Justus Heyer improved the method by which the increment was calculated during the conversion period and overcame the earlier problems through a laborious process of scheduling the way in which different parts of the forest were cut in different periods. Leopold Hufnagl, an Austrian forester in Slovenia and Herman Haupt Chapman in the United States also developed methods to include the increment in the calculations.

The various methods that were devised to control the yield from real forests could give markedly different results. In what is probably an extreme example, six methods were compared for an American forest. It was found that the cut that would be allowed by one method was twice the amount that would be allowed by another (Figure 4.2). However, the degree of difference between methods and their relative ranking could differ markedly in other forests. What was important for all the methods, was that the foresters re-measured the forests and then re-calculated the allowable cut every decade or so. Even if the methods were imperfect, the foresters could hope to bring their forests progressively nearer to their normal forest ideal when they would be able to cut what the forest grew each year, but no more for ever.

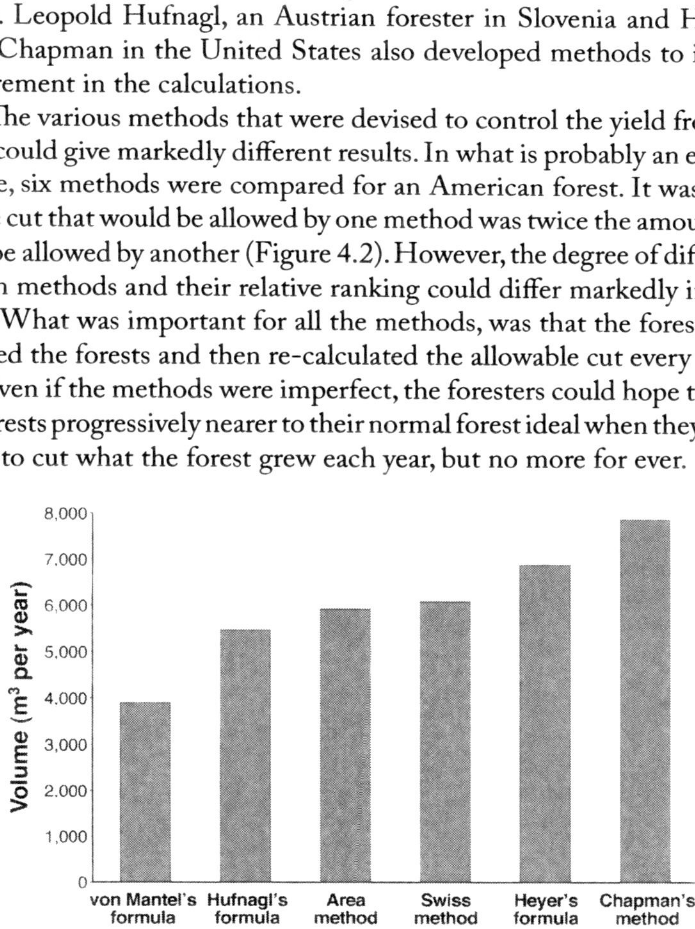

Figure 4.2. Annual allowable cut calculated by six different methods for a forest of red spruce in the Adirondack Mountains of the Eastern USA.

Source: From data in A.B. Recknagel *et al.* 1926. *Forest Management*. Original data expressed in cords indicating that the forest was probably used mainly for supplying pulpwood or fuel wood rather than saw logs.

The foresters' intense interest in developing better methods to bring real forests into the ideal 'normal' state illustrates the difficulty of the task for the even-aged forests. Their problem in selection forests was even more difficult.

Selection forests

Although the ideal of the normal forest was expounded for even-aged forests, its essential insight of balancing removal with growth was just as valid for forests whose stands kept a mixture of ages and species. The foresters still sought to increase the forests' productivity and gradually bring them towards an ideal state, but needed different methods. They had to exercise their control by selecting which *trees* to fell, rather than which *stands*. Two methods arose in the very different circumstances of Switzerland and Burma that set the ideas from which other methods were developed elsewhere.

The Swiss 'control method' was developed by Henri Biolley for communal forests in the Canton of Neuchâtel, as part of the growing reaction against the conversion of mixed forests by clear-cutting them and replanting them with even-aged stands of spruce.[6] To keep the Neuchâtel forests closer to natural systems, Biolley managed them with a silviculture which removed only selected scattered single trees or small groups at a time, so that a mixture of species, ages and sizes would always remain. Biolley's work was built on Adolphe Gurnaud's initial developments across the border in France. Their forests were particularly suitable for this selection silviculture as they were naturally composed of mixtures of shade-tolerant trees, like beech and silver fir, the light-demanding spruce, and several broad-leaved trees: oak, ash, linden, hornbeam and others. Moreover, some of them were on limestone 'karst' mountains that were particularly unsuited to clear-cutting as young trees failed to grow on the bare sites.

The scientific problem was how to devise a quantitative method for controlling the amount being cut so that productive capacity would be built up and sustained. The ideal of the normal forest had to be re-imagined. Rather than thinking of it as an orderly array of differently-aged stands, it had to be found in the way trees of different sizes grew naturally together: a few very large ones, more middle-sized ones and many small ones. After measuring many stands that were growing as naturally as possible, Biolley came up with an array of tree sizes – a distribution – that he thought was ideal. After analysing the distribution of a similar forest in the French Vosges mountains in 1898, François de Liocourt was able to provide a general

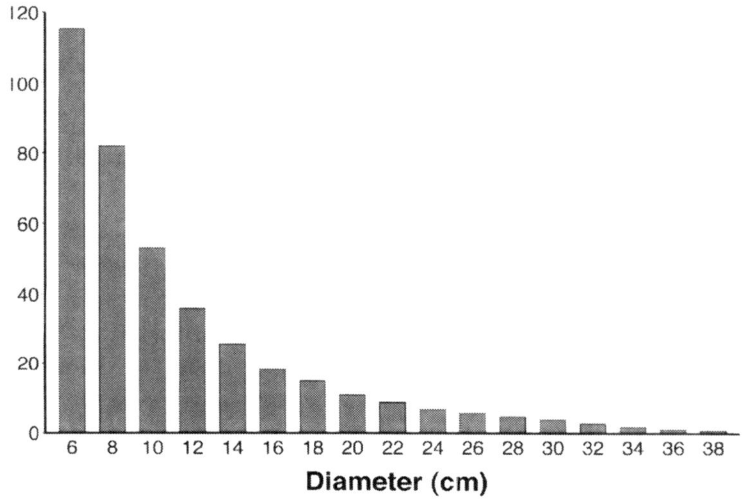

Figure 4.3. Biolley's distribution of tree diameters in an ideal forest.

Source: Plotted from data in Knuchel 1953, *Planning and Control in the Managed Forest.*

equation to describe ideal distributions.[7] This gave the method a universal quantitative foundation as its terms could be varied to suit forests in different regions of the world.

The method rested on controlling the number of trees in each size class. Only trees that were surplus to the ideal number for their class would be cut. To apply the method, the diameter of every tree in the forest had to be measured every ten years or so. When the distribution of the real forest was compared with the ideal one, it could be seen how many trees in which size classes could be cut. Rather than set a fixed conversion period, the process was one of progressive improvement. If the ideal distribution was realised the forest would be a normal one: the total growing stock would not change from year to year and the volume cut would equal the increment.

The problem in the teak forests of Burma with their mixture of ages and sizes was very different because only the valuable teak trees could be sold. A method of controlling a selection silviculture in the *jardinage* tradition was needed in order to regulate the yield from the whole forest. With very limited information, Dietrich Brandis developed a method of regulating the yield of the forest based on setting the number of trees over a set diameter that could be felled (discussed further in Chapter 7).

Limitations

The foresters' ability to realise their great hope of regulating the yields from the forests in order to sustain them in perpetuity was limited by their conception of the problems, an inherent contradiction with silviculture, the methods they had available and all too often by the overwhelming demands of industries with their political muscle.

The first conceptual problem was that the forests were never so readily reducible to being wood production farms as might be inferred from the way the foundation science was created. Even if wood was the most important commodity, the foresters also had to juggle it with hunting, grazing, litter collection and a host of other demands. The second conceptual problem was that the effects of their decisions outside the boundaries of the forest could be brought home to them, most obviously issues about soil and water by downstream users, and now by far-away interests in biodiversity or climate change. The forests were not entirely autonomous units and the foresters had to deal with various demands and neighbours in their every-day management.

Measurement, silviculture, economic analysis and regulation were the foundation stones of modern, scientific forestry, yet they were not as coherent as they seemed, or as the foresters desired. At least the foresters could reduce the limitations of their measurements with better instruments and sampling, but they could only gloss over their difficulties of trying to reconcile silviculture with regulation. Only in a normal forest were they compatible. If they chose the best silviculture for each stand in real forests, they did not get the steady yield they needed; if they regulated for a steady yield, they had to fell some stands too soon or too late. They could not find an exact solution because none was possible. It was not until 1980s that a formal method of reconciling silviculture with regulation became possible (as discussed in Chapter 10), and then only by accepting that the normal forest ideal would never be reached.

To exercise their control and regulate the amount of wood being cut, the foresters had to deal with these limitations. They did so by making formal 'working' or 'management' plans. These set out the owner's objectives for a forest, the information they had about it, how they calculated the allowable cut and where the felling, silvicultural and other work should be carried out. These plans were normally made for ten or twenty year periods, after which they had to be revised. In the process of making them, compromises had to be made that often had to rely as much upon judgement as calculation. Many countries gave the plans political authority by making their preparation mandatory for private owners, or by having them approved at high levels of government. Nevertheless, they were vulnerable to changing markets, political upsets or the ravages of storm and fire. What the foresters attempted, they could not always achieve.

Overarching all these limitations was the contradiction that lies at the heart of all conservation: unrestrained population and consumption cannot be sustained in a finite world. The foresters' great hope of eventually sustaining the forest's yield steadily and for ever was a static concept in an expanding world. It was not enough to increase the productivity of European forests and manage them systematically; forestry had to be extended to new places and new ways had to be found. The hopes and science of doing so are discussed in the next section.

Notes

1. Known as Area Regulation in English and *Flächenfachwerk* in German.
2. See Chapter 1. Evelyn, *Sylva*, ch. 32; *Diary* 3 Oct 1662 reporting a paper given to the Royal Society by Dr Christopher Merrett. Merrett was probably referring to the royal forests, Reed, pp. 44 –6.
3. German: *Massenfachwerk*.
4. German: *Normalwaldmodell.* Hundeshagen 1826.
5. The various methods are described in the standard forest management textbooks, e.g. Schlich 1895 or Recknagel *et al.* 1926.
6. French: *méthode du control.* Knuchel [1950] 1953; Johann 2006.
7. De Liocourt [1898] 2001.
8. Now the Justus Liebig University Giessen, named for its famous professor who developed agricultural chemistry and invented artificial fertiliser — discussed in Chapter 9.

EXTENSION

As described in the previous chapters, scientific forestry was founded in Europe to restore the productivity of the forests by applying the evolving methods of silviculture and regulation, or by planting as in Britain. The next three chapters show how it was extended to other countries and developed with other methods. The political and economic circumstances might differ from its European foundation, but the scientific base of rational enquiry, classification, survey, measurement, experiment and statistical analysis remained. Although the foresters had to devise new methods for different forests and found themselves no less busy with the tasks of every day, they did not lose their hope of sustaining the forests' yields for ever.

Forestry was part of the great nineteenth century expansion of European power that wove a net of science and influence stretching by longitude and latitude from the core imperial powers to their colonies, between the colonies and across to the New World. It was not only the ideas of forest science that were linked, but also trees. Their seeds were taken across the world so that pines from the northern hemisphere now support flourishing industries from Chile to South Africa, and Australian eucalypts do likewise from Brazil to Portugal. This story is told in Chapter 5.

The foresters sought to extend productive forests onto land that had not grown them before. They repeated a French example, allied to the restoration of sand dunes, around the world with variable success; and they established plantations using introduced species on land unsuited for agriculture. Their endeavours are discussed in Chapter 6.

The greatest extension of forestry was into the European empires. The foresters' desire to control the forests' yield matched the imperialists' control of the societies they conquered to create a distinct model in which forest science and practice evolved. Much of it was repeated in the United States, Canada, Australia and New Zealand, lands of new European settlement. But it was in the tropical rainforests of the colonies that the foresters faced their greatest challenges. Nothing in their European experience had prepared them for forests of such complexity, nor had they encountered such different societies. How they struggled to understand those forests and devise new forms of silviculture and regulation is described in Chapter 7.

5.

INTRODUCING

A type of forestry based on 'exotic' trees introduced from other lands was developed at much the same time as the foundations of scientific forestry based on the native 'indigenous' trees were being laid down. Its scientific origins lay in botany, as the splendour of the world's flora was being revealed by explorers and plant collectors penetrating ever deeper into Africa, the East, the New World and the Antipodes. They sent back increasing numbers of dried specimens, live plants and seeds to provide European botanists, gardeners and foresters with opportunities for research, delight and benefit.

The cultural origins of this approach can be traced back to the thirteenth century when the Holy Roman Emperor, Friedrich II imported numerous plants and shrubs from the Orient for his gardens and estates at Nurenburg in Germany. Botanical gardens evolved from such royal palace gardens, and from physic gardens growing medicinal plants, to become centres of research, education, public enjoyment and economic investigation. In the fourteenth century botanical gardens were established in Salerno and Venice; and in the sixteenth century at Prague, Königsberg, Leipzig, Breslau and Heidelberg. By the start of the nineteenth century there were major gardens at Kew in London, Schönbrunn in Vienna, Le Jardin des Plantes in Paris, Berlin and St Petersburg, and several smaller ones. At the end of the century there were well over four hundred, of which about 250 were in Europe and about 150 in the British and other European empires.[1]

In the seventeenth century it became fashionable to enrich European parks and gardens with introduced trees.[2] British estate owners attracted by their novelty ornamented their grounds with them as single specimens or as collections in an 'arboretum'. Such arboriculture flourished during the eighteenth and nineteenth centuries as a landscape and amenity movement in the tradition of John Evelyn, but the economic potential of the introduced species as forest trees only started to become recognised when their growth could be seen to be successful. Moreover, they only attracted economic interest in countries that lacked comparable native species, or where they offered some special timber properties.

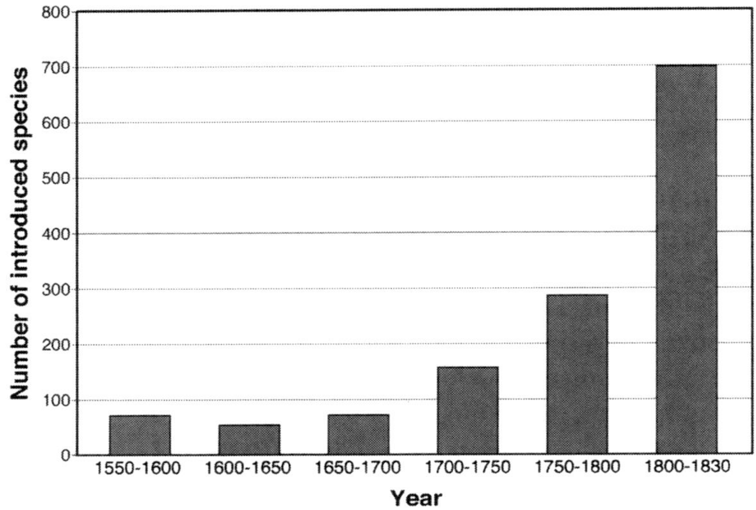

Figure 5.1. Introductions of foreign trees and shrubs into Britain showing rapid pace of increase from the mid-seventeenth century.

Source: Data derived from Loudon 1838. *Arboretum et Fructetum Britannicum; or the Trees and Shrubs of Britain.*

Collectors were despatched to scour the world for species new to science that would be attractive or useful. The pace of collection and the number of species introduced increased rapidly from the middle of the eighteenth century (Figure 5.1). By the end of the nineteenth century, trees from the west coast of North America, the Himalayas, Japan, China, South America and Australasia had been introduced to Europe and had become readily available to foresters around the world.[3]

Botanists had to name the flood of new plants arriving into the botanic gardens. They used the two-part Linnaean system of genera and species, but how best to separate the genera and how to name any species named differently by different collectors caused endless problems. It was not until 1867 when Alphonse de Candolle published a code at an International Botanical Congress in Paris that the botanists had a guide for resolving their problems and it was not until they held a Congress in Vienna in 1905 that they agreed to an international code.[4] At last the foresters could call on the international system for the trees' scientific names, but these were

Figure 5.2 Experiment with false acacia as a fast-growing species that could avoid a feared lack of timber.

Source: Frh. v. Schneider auf Negelsfuerst 1798. *Aufforderung zur allgemeinen Anpflanzung des unechten Akazienbaumes durch deren Anzucht dem Holzmangel bald dauerhaft abgeholfen werden könnte* [Call for the General Planting of False Acacia to Soon Permanently Remedy the Wood Shortage].

still subject to annoying changes as botanists revised genera and amended species names to fit the code.

In Germany the motivation for trying introduced species was a feared or existing lack of timber due to excessive cutting in the forests for charcoal. The forests needed to be restored and the wastes reforested as quickly as possible. Exotic fast-growing species promised to be a successful way for improvement. One of the earliest species to be tried was the false or common acacia that had been brought to Europe from America (where it is known as the black locust) in 1601, planted in Paris 1630, and introduced into Germany in the middle of the eighteenth century. Towards the end of the century, Friedrich von Wangenheim, a German botanist who had spent several years (1777–1785) in North America on military service advocated introducing trees from there. He recommended selecting species from climates similar to Germany's and pointed out the importance of 'provenance' in collecting seed only from the best strains of trees. He recommended red cedar, Douglas fir, eastern white pine and the common acacia, and made large experimental plantings of these and other species when he was appointed to a senior forestry position in East Prussia.

Scottish foresters led the way in establishing plantations of introduced species.[5] Scotland had the only native conifer species of commercial importance in Britain, Scots pine, but it was a limited resource and, like the rest of Britain, Scotland relied heavily on timber imported from Scandinavia, the Baltic and North America. The Scottish landscape had been extensively deforested over the centuries and the highland areas had been depopulated in the seventeenth and eighteenth centuries to create large estates whose economic base lay in sheep farming. This economy collapsed early in the nineteenth century as Australian wool captured their market. Tree plantations offered hope to those landowners wealthy enough to afford them. The most notable was the 'Planting Duke' of Atholl who had established over 3,200 hectares of European larch plantations by the late 1820s.[6] Later Japanese larch was planted which spontaneously crossed with the European larch to create a truly 'exotic' hybrid species in the 1890s.[7] Many other landowners established their own plantations for profit and pleasure. Such was the enthusiasm for introduced conifers that several of them planted their own 'pinetum' – an arboretum devoted solely to conifers.

Scottish forestry had four distinct features. First, it was rooted in the Scottish Enlightenment with its empirical approach to science. Second, it had an organised presence as the planting landowners and their foresters formed the Scottish Arboricultural Society in 1854, which added 'Royal' to its name in 1887 and published its own scientific journal, now *Scottish Forestry*. Third, education was highly valued in Scotland, the most literate country in Europe. The Arboricultural Society set up a formal education system through which forestry apprentices working on the estates could study at home to gain the Society's certificate. It was part of a culture that let able young men advance

Box 5.1. David Douglas (1799–1834)[8]

David Douglas introduced probably more species of trees into world forestry than anyone else. He was one of many Scots who made their way from humble starts as gardening or forestry apprentices on British estates to become leaders in forestry, horticulture and botany.

David Douglas, c1834.

Curtis's Botanical Magazine. Wikimedia Commons.

Douglas was born in Scone in 1799 as one of six children in a working class family. He started a seven-year apprenticeship as a gardener on the estate of the Earl of Mansfield at Scone Palace when he was ten years old. After this and working on another estate, he moved to the recently-founded Glasgow Botanical Gardens in 1820 when the outstanding botanist, W.J. Hooker was appointed as Professor of Botany in the University of Glasgow and the Garden's Director. Douglas could not have found a better lecturer and mentor, nor could Hooker have found a better assistant than this 'sturdy little Scot' of abstemious and pious disposition for their botanical expeditions through the Scottish Highlands.

In 1823 Douglas was commissioned by the London Horticultural Society to undertake arduous collecting expeditions. The first, short one was to the eastern United States and Canada. The second, for two years, was to the mountainous Pacific north-west region where he gathered the plants, birds and mammals that delighted his sponsors on his return. His third expedition, for five years, was to California and the Pacific North-west. He added over 240 species to British horticulture and thence to Europe and other parts of the world. Following the botany of the time, he placed all the eighteen conifers he collected in the genus *Pinus*. Some were later reclassified into other genera.

Douglas fir was the most important tree that he collected. He named it *Pinus douglassii*, but it is now *Psudeotsuga mensezii*. He wrote:

> Being an inhabitant of a country nearly in the same parallel of Latitude as Great Britain ... gives every reason to hope that it is well calculated to endure our climate and that it will provide a beautiful addition to the English Sylva if not an important addition to the number of useful timbers.[9]

His life as a great forest tree collector ended on the island of Hawaii when he was 35 years old, apparently by falling into a pit trap that also trapped a wild bull.

in careers that took some to lead forestry in the New World. Although they achieved a high standard of professional practice, the Society was well aware of the science education provided elsewhere in Europe and it lobbied for the forestry courses that were started in the Universities of Edinburgh and Aberdeen in 1889 and 1907.

The fourth feature of Scottish forestry was the energetic search for new species.[10] David Douglas was the most effective of a series of Scots collectors who went to North America from the 1790s. In the 1820s and 1830s, he introduced about 240 species new to British gardens and plantations, of which Douglas fir, sitka spruce, western red cedar, noble fir and grand fir proved successful as forest trees. Douglas also introduced monterey or 'radiata' pine, which did not thrive in Britain, although it later became an important forest tree in the southern hemisphere. The introduced trees proved so popular in Britain that they can now be glimpsed almost everywhere in the landscape, and they dominate the plantations. However, the British plantations remained of limited extent until 1920 when the state took a direct role.

The Scottish and English experience with introducing plantation trees was mirrored in other parts of Europe. For example, nurseries for raising seedlings of Douglas fir, eastern white pine, red oak and cedar were established in the forest owned by the city of Heidelberg from 1864 to1893.[11] German foresters took a scientific, experimental approach to introducing trees to their forests. In 1881 and 1885, the Association of German Forest Research Stations started systematic experiments with 45 Japanese and American tree species. In the course of time, Douglas fir, Japanese larch, eastern red oak, silver fir and sitka spruce proved to be of practical importance. Douglas fir was particularly important and in 1909 and 1910 several provenance experiments started with seeds from nineteen different sites and climatic zones to test variations in the form of the trees.[12]

Introducing trees beyond Europe

In the lands of new European settlement – such as South Africa, Australia, New Zealand, California and Brazil – plantations of introduced species were being established on a significant scale from the late nineteenth century. They were part of a flourishing acclimatisation culture that brought plants and animals as much for nostalgia as for economic benefit. It was oak, elm and ash trees, not the native eucalypts, that Australian settlers chose for their city streets and parks and they brought blackbirds for their sweet song and, to their eventual dismay when these became a plague, rabbits for the countryside. These species were only part of the long history of deliberately or accidentally moving plants, animals and diseases around the world.[13] Their success was uncertain. It was more likely if they came from similar climates, impossible if climates were grossly different, but possible with some species that could to adapt, or 'acclimatise', to different situations, an important matter as the world's climate changes in this millennium.

The economic benefit of introducing and disseminating trees was an important function of colonial botanical gardens.[14] Many of their botanists promoted forest conservation and tree planting and, in smaller colonies like Hong Kong and Mauritius, they were responsible for broad-scale tree planting. An outstanding example of their influence is the work of a German immigrant, Ferdinand Mueller (later von Mueller) who was appointed Victorian Government Botanist in 1853 and Director of the newly-established gardens in Melbourne four years later. As well as exploring, collecting and classifying plants and introducing many trees into the gardens, he urged the need for a forest culture. Impressed with the growth of radiata pine from Monterey in California, he despatched seed and young plants to be planted across the state.[15] Although radiata pine would eventually become the most widely planted tree in the world, it was in his enthusiasm for eucalypts that Mueller had the greatest effect on forestry in other countries.[16] They were already well-known. Seeds had been planted at Kew from 1774 and by the mid-nineteenth century eucalypts were growing in botanical gardens in Europe, South Africa and South America and a small plantation had been established in India. Mueller published *Eucalyptographia* (1879–1884), providing foresters with systematic descriptions and illustrations of 117 of the species. He knew there were twenty to thirty more, still to be illustrated. Now we know that there are over 700 species of which some 250 grow to the size of forest trees. He was a prolific correspondent who sent both information and packets of eucalypt seed to foresters in Europe, India, South Africa, Palestine, California and Brazil, foreseeing that 'eucalypts are destined to play a prominent part for all times in the sylvan culture of vast tracts of the globe'.[17]

By the end of the nineteenth century, foresters in the lands of new settlement could see that introduced trees offered them many opportunities: some species grew fast or to an enormous size; the conifers produced the softwood timber keenly demanded in countries with little of their own; the eucalypts produced wood for fuel, paper, timber and railways; some acacias had bark used for tanning leather; teak was a fine timber; and so forth. They needed to move beyond botany: with so many species becoming available, they needed a systematic, quantitative way of deciding which to plant. They did so in what ideally was a four-staged process of species introduction, experimental arboreta, trial plantations and plantation review. There was never enough time in practice. Even the fast-growing introduced trees took twenty or thirty years before they could be evaluated, so that extensive planting often had to be started before the early stages had been completed.

Rather than rely on trees the botanists happened to have collected, the foresters needed a better way of finding species that were likely to do well, and they needed their own experimental arboreta where they could grow and measure plots of 50–100 trees of each promising species. Then they could start trial plantations to see how those that did best in the experimental arboreta fared in stand conditions of a few hectares. Finally they could select the species for full-scale planting. However, as full-scale planting extended onto sites they had not tried, the trees might not thrive, the expected growth rates might not occur or the timber quality might be poor or unfamiliar in the market. They needed to review the success of entire projects once significant areas had grown to maturity.

The course of these stages can be followed in South Africa.[18] Small plantations of eucalypts and pines had been started in the 1880s to augment the dwindling wood supplies from the slow-growing native trees, but from the start of the twentieth century more wood was demanded for mining and railways and the rate of planting was increased. David Ernest Hutchins, an experienced English forester who had been trained in the French forestry school and had worked in India and the Cape Province, led the scientific development with a study of the Transvaal Province in 1903. Notably, some plantations that had been started by mining companies there had failed. Hutchins knew that 'every tree has its own habitat, its own peculiar air and temperature, and it is little sense attempting to plant it out of its habitat'.[19] Species should be selected from similar climatic regions, or 'homoclimes', which meant that, for the inland plateau of the Transvaal with its largely summer rainfall, they should be sought from Mexico, the drier Western Himalayas and part of the Argentine. He also thought that eucalypt species from the southerly latitudes of inland eastern Australia were worth testing.

As government plantations were started in various parts of the country, Hutchins insisted that experimental arboreta be included. This enabled species to be tested across climatic differences as well as compared in each zone. One of the areas where trees from Mexico, China, Japan and Malawi were tested was at Woodbush where a young forester, Charles Edward Lane Poole was in charge. He carried Hutchins' ideas with him when he took charge of the Forests Department in Western Australia in 1916.[20] There, he planned a series of seven arboreta sited from hot dry Kalgoorlie to temperate Nannup. As well as testing pines and cypresses for conifer plantations, and eucalypts from other parts of Australia, the arboreta were expected to 'add beauty to the township'. When Lane Poole moved to the Australian Capital Territory in 1927, he started a more ambitious series of 37 arboreta designed to test 242 species from 37 countries across a range of elevations and climates.[21] Similar series of experimental arboreta and small trial plantations were tried in other parts of Australia and in other countries.[22] The best conifer species to use for full-scale plantations had usually become apparent in the early stages, but choosing the best eucalypt species was more difficult as many are not easily identified when grown away from their native land, or when the original seed source is unclear.

The fourth stage of reviewing whole plantations was usually called for when parts of them had failed to live up to expectations. The reviews required quantitative assessments and investigations of the reasons for their performance. In South Africa, for example, Lane Poole was called back in 1935 to review 20,000 hectares of state eucalypt plantations that were no longer required to supply wood to the railways. He found that their wood was of poor quality due to the way the trees had been grown and that pines would have been done better. A long review in South Australia in the 1930s was particularly important because the state had few natural forests and had established radiata pine plantations to supply its own wood. They had been advocated by Moritz Richard Schomburgk, the German-born Curator of the Adelaide Botanic Gardens, in 1875 and started by a Scottish forester, John Ednie Brown. By the 1930s it was clear that unless the stands could be thinned and pruned the quality of the wood would be poor and the enterprise unprofitable. A review by an Australian forester, Edward Harold Fulcher Swain found that 8,000 of the 19,000 hectares were ineffective because they had been planted on poor sites. Fortunately the rest grew well, a zinc deficiency was corrected, a pulpwood market was found that enabled the stands to be thinned and planting was continued. Today, South Australia has a fully-developed forest industry based on thriving plantations.

The staged process of botanical collection, arboreta experiments, trial planting and project evaluation resulted in only about twenty out of the thousands of forest trees in the world being selected for fast-growing exotic plantations on an industrial scale. Most of these had been clearly recognised by the 1920s and 1930s. Of the conifers, the outstanding successes were: radiata pine in Australia, New Zealand and Chile; pinus patula in South Africa; and Sitka and Norway spruce in Britain. Of the eucalypts: they were blue gum from Tasmania in Ethiopia and other countries; Sydney blue gum – a closely related tree from a warmer region – in Brazil, Costa Rica and Uganda; and river red gum in Israel and Turkey. As foresters in tropical countries sought to develop industrial-scale plantations from the 1950s, they sought trees from other tropical regions and found two that proved remarkably successful: hickory wattle or mangium from North Queensland and Papua New Guinea for plantations in Indonesia, Vietnam, Malaysia; and gmelina from the Indian sub-continent and South-east Asia for Brazil, West Africa and Malaysia.

The history of forest tree introductions outlined in this chapter reveals the great hopes that foresters carried for improving the wood yields from the land and the mix of sciences and practices that they brought to the task from botany, climate analysis, experiment and measurement, to trial and evaluation. Compared to agriculture, it was only the first step in the journey of domesticating wild plants for human use. The steps of breeding and intensive cropping would follow as described later in Chapters 8 and 9.

Notes

1. McCracken 1997.
2. Introduced species were well known in France (Versailles, St. Cloud and Fontainebleau) in the sixteenth and seventeenth centuries and in Germany in the eighteenth century (1750 Harbke near Helmstedt, 1768 and 1808 Wörlitz).
3. Campbell-Culver 2001.
4. Nicolson 1991.
5. Anderson 1967. Oosthoek 2013.
6. Anon. 1832.
7. Edwards 1956.
8. Information drawn from Piggott 2004; Douglas 1914.
9. Douglas 1914, p. 340.
10. House and Dingwell 2004.
11. Bendix 2008.
12. Heyder 1983.
13. Crosby 1986 has brought out the extent and consequences of introducing European organisms to the New World, but this is only part of the story, since organisms travelled

to Europe and laterally across the world, as in the case of introducing Pará rubber trees from Brazil to Asia in the 1870s.

14. McCracken 1997.
15. Morris 1974; Gillbank 1993.
16. Doughty 2000; Zacharin 1978.
17. Cited by Zacharin 1978, p. 58.
18. Grut 1965; Hutchins 1905; Darrow 1977.
19. Cited by Darrow, p.10.
20. Dargavel 2008.
21. Shirley 2008.
22. Gaughwin 1999.

6.

CONVERTING

Not only could better trees be introduced from other lands, as described in the previous chapter, and the existing forests tended scientifically, but the European foresters could also make them more productive by converting them in two radical ways. One was to make better use of the trees they already had by growing their conifers instead of their broadleaved trees. It was not just a matter of favouring the conifers over the hardwoods where they grew together, but of planting conifers where they had not grown before. In this sense they were introduced, but they were also familiar. It was a general process in all parts of Central Europe as the foresters converted large swathes of the broadleaved forests with their mixture of species to conifer forests of single species. The other way was to engineer the land so that it could grow trees where they had not grown, or not grown well before. These conversions radically changed the landscapes of Central Europe, France and Scotland, as discussed in this chapter, to much of what we see today.

Conifer conversion in Central Europe[1]

Conifers are so prevalent in the landscapes of Central Europe that it is hard to imagine that they have not always been thus. However, scientists have analysed the pollen preserved for centuries in the mud of ancient lakes and swamps and have been able to reveal the extent and greater diversity of the original Central European forests. Although conditions varied widely, broadleaved forests with many different mixtures of species, including conifers in places, predominated. They were certainly more complex than those that remain. In some forests relative proportions of conifers and broadleaved trees were altered, while in others the original species were completely replaced by planting. Four partly interrelated forces drove the changes in different periods: economics, ecology, politics and management. Some changed the relative proportions of conifers and broadleaved trees, others artificially regenerated the forests to change their species. The period from the Middle Ages to the beginning of the nineteenth century is characterised by the intense use of the forests and sometimes by their heavy exploitation. It varied by time and region. On the one hand, forests were cleared for arable land from the fourteenth century – as they had been in Western and Southern Europe for three thousand years. Some countries left themselves with very little forest; Scotland had only four per cent by 1750 and Denmark only two or three per cent by 1800, for example.[2] On the other hand, forests were used intensively to supply wood for mining and industries in ways which progressively degraded them, changing their density, structure and species composition. It was these situations that the foundations of scientific forestry, discussed in earlier chapters, had sought to correct.

Although some of the changes that increased the proportion of conifers over the centuries resulted from grazing, browsing by game, selective cutting or harvesting the litter, the systematic conversion of the broadleaved forests by introducing conifers dates from the middle of the eighteenth century in many parts of Germany (Mecklenburg, Brandenburg, Saxony and Bavaria).[3] There were several reasons for favouring conifers. Many previous attempts to restore degraded broadleaved forests by clear-cutting and regenerating them naturally had been unsuccessful, often because the young trees were suppressed by weeds.[4] By contrast, planting Norway spruce suppressed the weeds and any young beech and fir trees that appeared, leaving a 'pure' stand of spruce. The conifer seedlings were easier to grow in the nurseries than the broadleaved trees. Economically, the demand for spruce and pine softwood was high, the rotation period to grow them was short, the small trees removed in intermediate thinning could be sold and the final cutting yielded plenty of wood. Conifers were also more suited for planting in waste land with poor soil. Moreover, the classic schools of German forestry where the foundations of modern forestry were laid favoured conversions. Norway spruce and Scots pine were the preferred species with larch, eastern white pine and Douglas fir also being used later.

Conversions increased both the areas of European forests and their productivity. The areas were increased in almost every European country, since their lowest point in about 1800.[5] Much of the increase in productivity was due to the conversions. However, they created typically even-aged monocultures of relatively young stands whose productivity in Saxony, Bavaria, Silesia, Bohemia, Prussia and in the Austrian Alps has been reduced by increased wind, snow and insect damage. Although the foresters adapted their silvicultural methods over the course of time to minimise these risks, climatic change is likely to increase it. This and the environmental effects are discussed in later chapters. In this chapter we discuss three cases of conversion in the Karst region of what is now mostly in Croatia and Slovenia, Les Landes de Gascogne in France and the Scottish moors.

Greening of the Karst region

There are many regions of karst landscapes in the world, but the one discussed here lies between the Alps and the Mediterranean. In the nineteenth century it was part of the Austrian Empire but is now in Croatia and Slovenia. It takes its name from its characteristically dry landscape of limestone and dolomite rocks, with their sink holes, caves and underground rivers. It is harsh country and over 400,000 hectares had become so severely degraded by the early nineteenth century that the state decided to reforest one-third of it.

For centuries the region had been used for grazing sheep, cattle and goats, but overgrazing, mainly by the goats – 'the cows of the poor' – was largely responsible for the degradation. The oak woodland had been grazed, the best grasses had been lost and the soil exposed to wind and sun. When its humus had decomposed, the plant roots and tree seedlings had died and every downpour of rain had washed more soil into the valleys, leaving only the naked rock. There had been forest on private land that was reserved from felling until controls were abolished in the early nineteenth century; the wood traders moved in to supply the French and British navies and the goats followed. Deforestation had been so relentless that by 1850 only five to ten per cent of the land still had trees. The degradation of the landscape had impoverished the population, led many farmers to emigrate and caused the state to lose tax revenue.

Reforestation was led by a talented but unusual forester, Josef Ressel. He started in the early 1820s and made his first attempts to reforest the island of Veglia (now Krk) in 1837. With that experience behind him, he presented a well-designed plan for the reforesting the hinterland of Trieste and Goricia in 1842. If Ressel was to convert the land to productive use again, he had to find ways to overcome problems that were both technically and socially very difficult. He developed several methods for getting trees back onto the land. One was the arduous work of carrying soil in wicker baskets to the holes where the trees were to be planted. Another was planting hardy shrubs of common juniper so that five years later oak and larch could be sown underneath their protecting cover and another was planting the hardy Austrian pine.

If his methods were to succeed, Ressel had to persuade the peasants, long used to grazing their goats and other animals across the landscape, that

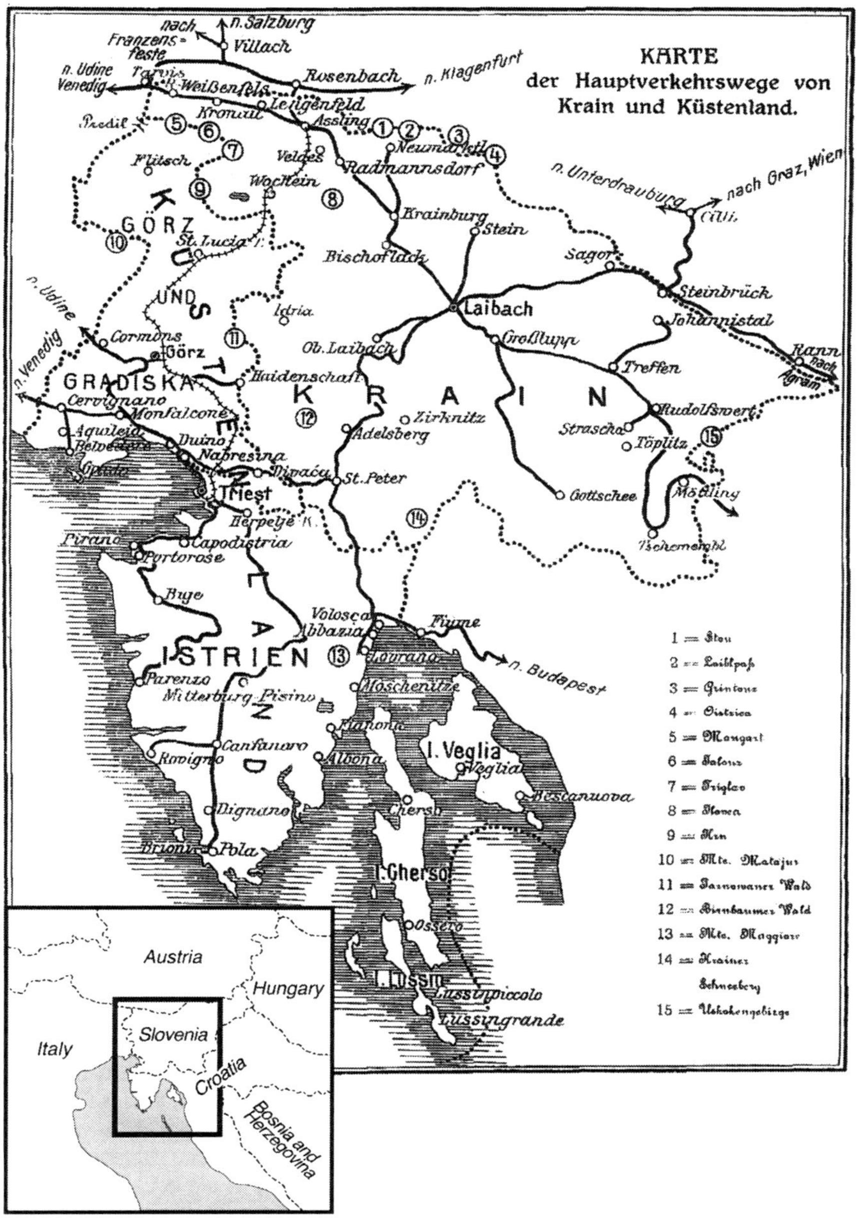

Figure 6.1. Map of the Karst region.

Source: *Verkehrsbuch österreichischer Eisenbahen* [Trafficbook of the Austrian Railways] 1910.

Box 6.1. Joseph Ludwig Franz Ressel (1793–1857)

Josef Ressel

Portrait from the Dibner Library of the History of Science and Technology, Wikimedia Commons.

Joseph Ressel was born in 1793 in the east Bohemian town of Chrudim, now in the Czech Republic. His father, Anton Hermann Ressel, was German-speaking and his mother, Marija Ana Ressel (née Konvichkova), Czech-speaking.

In 1806 Ressel left Chrudim and entered the grammar school in Linz and in 1809 he began a two-year study of artillery at the 4th Artillery Regiment in České Budějovice (Budweis) where he acquired an outstanding knowledge of mathematics, geometry and technical drawing. He went on to the University of Vienna from 1812 to 1814 where he studied state accountancy, chemistry, veterinary medicine, agriculture, forestry, hydraulics, architecture and natural sciences. When his parents were no longer able to support his studies he left university and continued studying for another two years at the newly founded Forestry Academy at Mariabrunn near Vienna.

In 1817 he was appointed district forester in Pleterje/Carniolia and later a forester in Motovun, Trieste and Ljubljana. From 1837 onwards be became a forestry official for the Austrian navy, responsible for supplying a sufficient quality and quantity of ship building timber to the Navy's shipyard in Venice. He was married twice, first to Jakomina Orebich from Motovun and, after her death, to Tereza Kastelec from Višnja Gora. Ressel died in Ljubljana in 1857.

He was an inventor and a forester of high skills and creativity. His most famous invention was the permanent screw for steam ships, but there were many others. As a forester he became well known for his originality in supplying timber to the Austrian navy. His program included long-term agreements with the farmers, unorthodox forest management systems and socially-adjusted planting programs. His vision of greening the Karst seemed out of reach in his time, but was realised within the lifetime of the trees.

reforestation would benefit them. This was not only a practical matter for Ressel; he believed that local forestry administrations should take responsibility for reducing poverty. Now we might see his work as a precursor to the social forestry movement a century and a half later (Chapter 13). He developed several strands to his socially-oriented approach by working out how to reforest the common lands in Istria.

Ressel fought against the contemporary scientific theory and practice of growing trees in dense stands. Instead he kept them widely spaced to meet the shipyards' demands for curved timber on one hand and the peasants' demands to have grass under the trees on the other. He acknowledged that his ideas were opposed to the classical school of forestry of his time but defended his plan by demonstrating practical examples. He talked to the farmers, included their demands in his scheme and explained the details to them. He proposed planting forest trees such as oak, elm, mountain ash, lime and others on the mountain sites exposed to the wind and cultivating nut trees (hazelnuts, chestnuts and almonds) and fruit trees (plums and pears) in protected places. Ressel started on a small scale by teaching the young people in their mother tongue how to establish nurseries, plant trees and take care of them. At the same time the schoolchildren started to plant juniper bushes.

Having started on a small scale, he moved on to a larger scale in the forest districts of Studenice and Motovun where he arranged for plants to be raised from cuttings. All the activities were based on his proposals; for example, chestnuts were planted in the lowlands and nut trees the higher country. Easy-readable textbooks were distributed, which also were used in schools. Training centres were established in Sezana, Prosecco and Goricia to teach capable people how to cultivate fruit trees. The workshops lasted three to five days and the journey to reach them was financially supported. The results were satisfying and gradually stimulated rural people and church representatives to plant trees. Their efforts were assisted with seeds distributed by the government to the local farmers, while the communities provided the oak and chestnut seedlings and in some places the teachers grew the plants.[6] From the 1860s reforestation to convert the degraded landscapes expanded to the regional scale.

It was not until fifteen years after Ressel's death in 1871 that a special forester was designated by the government to guide similar activities in other regions. However, the forestry theory at the time favoured conifers, so that some of the broad-leaved forests that Ressel's scheme had fostered were replaced by spruce in several places. In the following decades black pine, the most successful of the species, was spread all over the karst region.

Ressel was aware that his ideas were not welcomed by the other foresters, who aimed for dense monocultures of conifers to bring the highest profit and who more or less condemned growing broad-leafed trees. The disagreement among his colleagues and the official forest administration in Vienna was so strong that after Ressel's death, the office threw away all the plans he had made during his working life. Fifty years later some foresters who admired his work wanted to celebrate his life. They looked for copies of his plans in places where he had worked, in Venice, Trieste and Ljubljana, and found some manuscripts that had survived. They are stored now in the Technical Museum of Vienna and are highly appreciated today, particularly in Slovenia, which acknowledges that having half its land covered by forests can be traced back to the vision of a forester with an ambitious scientific interest and a sense of social responsibility.

Les Landes de Gascogne

Perhaps productivity could be increased if the very land itself could be made more suitable for trees, or so the foresters hoped. It was as much a matter of engineering and technology as of science, although much science was needed to find the causes of success or failure. In converting the land to its needs, forestry was only following the long history of agriculture and joining the eighteenth century drive to 'improve' the natural world. Drainage, cultivation and privatisation by the 'enclosure' of common land were its means.

The French case became the exemplar taught in forestry schools, described in textbooks and imitated around the world. It was set in the Les Landes de Gascogne area in the southwest of the country, where the effects of stabilising drifting coastal sand dunes and draining swamps could be proudly displayed.[7] What had been France's poorest region, plagued by malaria and pellagra and subsisting by growing meagre crops of cereals or grazing sheep on the open moors, had been converted into pine plantations. Their wood and resin supported a prosperous forest-based economy that was developed in the last quarter of the nineteenth century. Although this 'Golden Age of the Forest of Les Landes' declined in the 1920s, it was re-invigorated by industrialisation. It was not only an example of how land could be improved for forestry, but also the world's first example of how large plantations could stimulate the development of a whole region.

The problem of how to stabilise the coastal dunes to prevent their sand drifting inland was particularly acute. The area was a large, low-lying triangular plain of about one million hectares that fronted the Atlantic with Europe's largest dunes. The land was nowhere more than about one hundred metres above sea level and, over time, the movement of dune sand inland had engulfed whole villages. A civil engineer, Nicolas Thomas Brémontier devised the first successful system to halt it in the 1780s. His system consisted of building a double palisade or brush fence inland from the high tide mark. When a dune had started to form above the fence, it was stabilised on the ocean side by planting marram grass. On the lee side, it was stabilised by sowing seeds of shrubs and the native maritime pine, protected by a covering of brush and branches. Although the growing trees became stunted and were often half-covered by sand, they trapped it effectively. Practical experience with the system enabled the best height and slopes for the dune to be specified.

The problem of how to improve the moors behind the coast was tackled by another engineer, Jules Chambrelent.[8] He was sent in 1837, when he was only twenty years old, to investigate the poor quality of the land and see what kind of products it might yield. It must have seemed a hopeless task: the land was parched in summer and broken by flooded swamps in winter; several private drainage and agricultural development companies had failed, as had attempts to grow various crops. When he arrived, he found much of the area to be almost uninhabited, with only isolated thatched cottages whose people had to use stilts to cross the swamps. It was an abrupt contrast to the prosperous city of Bordeaux and its surrounding vineyards on the area's northern border. Undeterred, Chambrelent tackled the task in a quantitative and scientific manner by making a detailed topographical and soil survey covering 800,000 hectares. It took him five years.

Chambrelent took great pains with the accuracy with which he measured the height of the land and was a thoughtful observer of the area's general conditions. He reported that the very slight slopes of the land, about one metre per kilometre, lay in the same direction. Natural irregularities on the slopes were less than half a metre in height yet they were responsible for the flooding problems, because an impervious iron pan lay forty to fifty centimetres under the surface of the sandy soil. He devised a simple scheme of shallow ditches that could drain the land by cutting through the irregularities. He noted that the occasional clumps of pine trees grew well on any well drained rises and concluded that they would be the most suitable crop for most of the area; the soil was too sandy for much else. Chambrelent's scheme was not accepted; perhaps it seemed too simple and a lone young engineer too naive when compared to a large agricultural development scheme directed by 'the most distinguished people and richest capitalists' that had been started at the same time as his survey.

Chambrelent was so convinced of the soundness of his scheme that in the summer of 1849 he acquired 500 hectares in the most difficult part of the moors and tested it at his own expense. He surveyed the gradients of the ditches very carefully to ensure that the water could trickle away, and he sowed pine and oak seed during the next spring. The results were remarkable: the winter rains drained away, the seeds germinated in the moist, sandy soil and within five years he had trees 3.5 metres tall. By then other landowners had followed his example and converted another 20,000 hectares. Meanwhile, the large development scheme had failed to produce any results from its investments. Recognition for Chambrelent came from that particularly nineteenth century institution, the international exhibition. Not to be outdone by Britain's Great Exhibition of the Works of All Nations in 1851, France held its own Exposition Universelle in 1855. It emphasised agriculture, enabling Chambrelent's scheme to be entered. This not only won a gold medal, but the international judging panel recommended that it should be applied on a far larger scale; and for that to happen, the state had to be involved.

To apply the scheme on a larger scale, main channels would have to be dug to collect the water drained from the moors and roads would be needed for access. For this to be worthwhile, the state had to ensure that it would be taken up by the local people. In the era of *laissez faire*, the state would not intrude on private property, but it would on the commons. About one third of the area was held as commons controlled by the local governments, the 'communes'. In 1857 the state legislated to compel them to drain and sow trees on their moors at their own expense. However, it funded the construction of rural roads with collector drains and a major outlet to take the water to the sea.

Over the next twenty years the region was transformed because the physical conversion of the land was closely linked to economic and social changes. The communes could only afford to drain the commons by selling them. By 1877 they had sold two thirds of their commons, the sheep owners had lost their free grazing and a new class of owners had been brought into the region. Not surprisingly, it encountered fierce opposition from the shepherds or *pasteurs* whose meagre livelihoods depended on their access to the moors.[9] The hated pine plantations could be burnt and in 1870 and 1871 forest fires destroyed over 50,000 hectares of the stands created under the 1857 legislation. Arson can rarely be proved and this area where much of the local population was sympathetic was no exception.

Modernisation also brought benefits. As the improved land became more valuable, the sales enabled the communes to build schools, churches

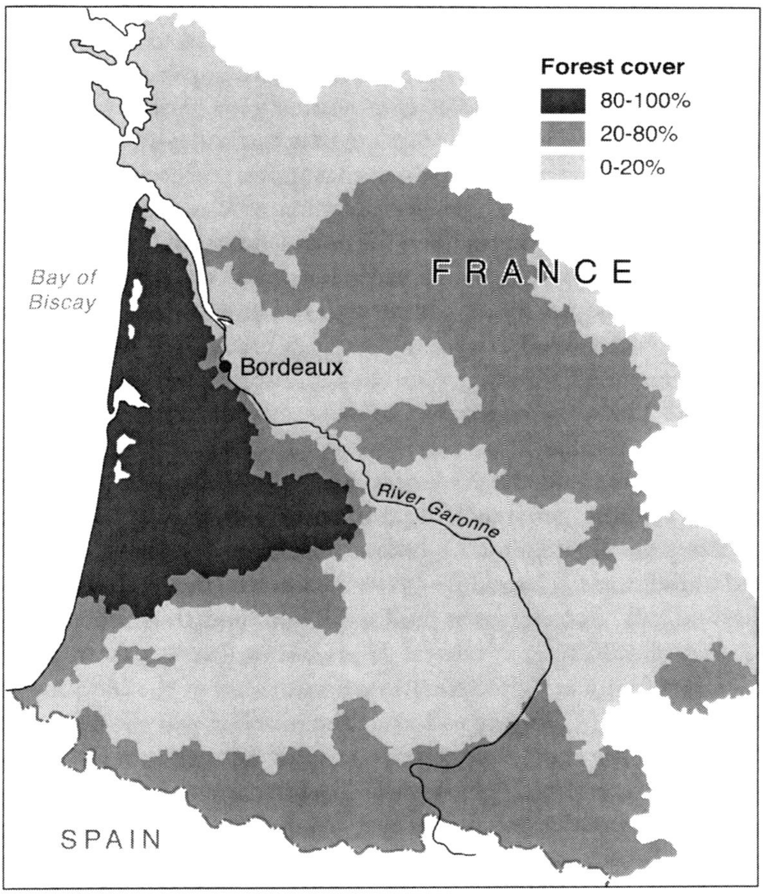

Figure 6.2. Les Landes de Gascogne showing forest cover by 1950s.

Source: Redrawn from Barrère and Lerat 1962. *La Règion du Sud–Ouest.*

and presbyteries as well as drain more land. The health of the population improved, partly because Chambrelent developed a system of filter wells that provided clean drinking water. The process accelerated as other owners started to convert their own land once they saw how successful the new plantations were. The combined effects turned Les Landes into an important forest region and the old agro–pastoral society gradually gave way to the new modern order, geared to supplying industries (Figure 6.2).

Initially, the region's expanding economy was largely based on exporting its forest products. Historically, maritime pine along the coast from France to Spain had been tapped for its resin, used to make the turpentine and marine tar – 'naval stores' – for wooden ships and to provide the base for paint and many other uses. As the plantations expanded, tapping resin provided work for about 20,000 men by the 1920s. Although the market for resin slumped, the wood was exported far and wide: to Greece, Algeria, Tunisia and Spain; pit props for British coal mines; and even railway sleepers for Uruguay and telegraph poles for the Panama Canal. Domestically, the plantations supplied French markets for timber, firewood, vineyard posts and wooden blocks to pave Paris boulevards, but it was the paper industry that secured their economic future. Faced with the loss of their resin market, a group of owners founded a paper mill in 1925 to use their plantations. Five industrial companies followed their lead and set up paper and fibreboard mills. By the 1960s the region had become a significant centre for manufacturing forest products.[10]

Elsewhere along the European and later American coasts, similar problems of stabilising coastal dunes and converting poor heath land to forests were addressed.[11] In Denmark's Jutland region, for example, the foresters planted a forest belt of introduced trees behind the dunes that enabled the land behind it to be converted to productive farmland. Belgium followed the French way of stabilising its coastal dunes, but its foresters faced a different drainage problem when they tried to convert some of the high moor land on the border with Germany to forest. The problem was the peat: the only soil into which they could plant was a layer of wet peat: even if the young trees survived having 'wet feet', they would hardly grow. The Belgian foresters devised a system of digging drains every four metres across the moors and using the turfs they removed to create mounds which dried out in enough in a year for planting the trees. It was the example that was to change the face of British forestry.

Scottish moorlands[12]

Sir John Stirling-Maxwell, a wealthy Scottish estate owner, realised the importance of the Belgians' system when he studied it in 1905 during a tour with the English Arboricultural Society. His foresters had been experimenting with different methods of planting trees on his own high moors but with little success and he hoped that the Belgian drainage system would enable his land to be 're-peopled with trees and busiest with the healthiest of industries'. They conducted more experiments, added fertiliser as well as digging drains and showed that trees could be grown on peat-covered mountain land. The small-scale experiments of an enlightened owner were not enough at the national scale, as Britain belatedly realised when in 1912 it started a national survey to find land for planting. Only on the high, almost treeless moors was a significant area of land available and much of that was covered in peat and heath. Roy Robinson, a graduate of the new forestry school at Oxford University, was appointed as one of the two researchers for the survey. He selected a demonstration area in the west of Scotland to try what could be done. He needed to estimate the relative quality of different parts of the area, but without trees to measure he could not determine it from their heights; he could only use the soil, grasses, herbs and heaths. He surveyed this vegetation and started experiments to find the best species to plant and the best way of doing so.

These small beginnings were not enough to address the national problem that became starkly apparent during World War I. Britain was too reliant on imported timber, did not have enough forests and many were in poor condition, even before wartime felling. The long history of deforestation had left Britain as one of the least forested countries in Europe; once forests had covered three quarters of its land but by the time of the War they had been reduced to five per cent, mostly in small separated patches of woodland in agricultural landscapes.[13] Coal, not wood, fuelled Britain's industries and its timber came from the Baltic, North America and countries around the world. Forestry was better established in its Empire than at home: something had to be done.

In 1919 Britain established its Forestry Commission to restore the nation's limited forests and establish extensive plantations. It would have to convert much of the moors and plant any other type of available land. It had to move beyond the demonstration areas and estate forestry and develop the scientific basis for assessing the existing forests and the land available across the country. It was a complex problem because the forests and land had been highly modified over the centuries and both the introduced and the native trees had to be considered. With Stirling–Maxwell and Robinson as two of its first commissioners, it had a firm commitment to establishing the work on a scientific basis.

The emerging science of plant ecology provided a foundation on which the assessment method was developed by British scientists. They had been stimulated by the work of Eugen Warming in Denmark and Andreas Schimper in Germany in the 1890s and had started to survey the vegetation of different parts of the country. They had formed the British Ecological Society in 1913 and published their findings in its journal. Their work provided foresters with classifications of plant communities and large-scale vegetation maps of broad areas. However, the foresters needed closer work because their areas were highly variable; geology, soil, climate, topography all affected the growth of trees differently according to species. Moreover, the ecologists' broad classifications based on natural or semi-natural conditions were not applicable on the moors, where their general term 'peat' covered a wide range of soil types.

Mark Anderson, in the Commission's small research branch and later in Ireland and the universities of Oxford and Edinburgh, put the assessment of planting sites through vegetation onto a scientific basis in publications from 1932 to 1961.[14] His work was based on trying to detect the type of forest likely to have been present historically. As the land had been so modified over the centuries and largely deforested, he had to rely on close inspection of the remaining vegetation, often finding the grasses and mosses to be the key indicators. He was particularly interested in plantations that had failed because the particular features of their sites had not been allowed for when the species were selected. His prime concern was that the right species should be planted in each part of an area. In work oddly reminiscent of Evelyn's three centuries before, Anderson detailed the characteristics, site requirements, uses and ways of growing each of the major possible plantation species. Whereas Evelyn had only discussed seventeen, Anderson discussed 48 in order to include the introduced species. He classified them and noted whether they were 'hardy' enough to withstand winter frosts, or were 'tender' and would require the shelter of bigger trees on exposed sites when young.

Anderson's method was designed for practising foresters. It required the close examination of every site to identify its plant community, using indicator species in modified areas if necessary. He classified the communities into twenty classes by their fertility and moisture (Table 6.1) and then matched the species to the communities (Table 6.2). His simple guide could be modified to suit particular circumstances or species requirements. His selection was ruled by the need to maintain or improve the soil fertility; only when two species would grow equally well did he think that the choice might be made on economic considerations. Most of the land available for planting was in the lower fertility classes suited primarily to Sitka and Norway spruce on the wetter sites and Scots pine on the dry or moist sites.

Table 6.1. Anderson's classification of plant communities on land for plantations.

Fertility class	Moisture class			
	1. Dry	2. Moist	3a Wet	3b Wet with peat
A	Dry grass–herb	Moist grass–herb	Grass–rush (Hard rush)	Willow–reed
B	Dry–grass	Fern	Sedge	Soft rush
C	Grass–heath	Rush–grass (Jointed rush)	–	*Molina*
D	*Erica cinerea*	*Nardus–Molinia*	–	Cotton grass
E	*Calluna*–heath	*Vaccinium*	–	*Myrica*
F	Lichen	*Erica tetralix*	–	*Sphagnum* – moor

Table 6.2. Three examples of Anderson's selection of tree species for sites.

Fertility class	Moisture class	Hardy species	Tender species
B	1. Dry	Elm, Lime, Whitebeam, Cordate alder, Austrian pine, Corsican pine	Ash, Beech, Sycamore, Horse chestnut
C	2. Moist	Birch, Aspen, Alders, Scots pine, Thuja, Lawson's cypress	Oak, Spruce, Tsuga, Silver firs
D	3. Wet	Pubescent birch, Alders	Sitka spruce

Some of Anderson's early experiments investigated ploughing the moors to break their iron pan layers and drain them. Powerful tractors that became available after World War II enabled this to be done effectively and allowed the moors to be converted to plantations more rapidly. The physical conversion

Box 6.2. Mark Louden Anderson (1895–1961)

Professor Mark Louden Anderson

Photo from Anderson (ed. Taylor) *History of Scottish Forestry*, 1976.

Mark Louden Anderson was born into a religious family: his father, uncle and grandfather were ministers in the Church of Scotland who imbued his life with the principles of hard work, frugality, efficiency and integrity. His parents died when he was twelve years old and he was brought up in his uncle's family. He entered Edinburgh University in 1912 to study forestry, but broke his course to join the army in 1914. He served in France and was awarded the Military Cross. On return, he re-entered the university and completed his degree.

He became a Research Officer with the Forestry Commission 1919 and was awarded the first doctorate in forestry from the University of Edinburgh in 1924 for his work. He moved to the Republic of Ireland in 1931 as Chief Inspector and Director of Forestry. In 1946 he moved to the Imperial Forestry Institute in Oxford University and in 1951 became the Professor of Forestry in Edinburgh University. He was a controversial figure, as his passion for forest science was accompanied by intolerance for those who did not share his principles and his insistence on basing forestry on ecology often put him at odds with the widespread practice of planting conifer monocultures – 'nothing but cabbage planting here' was his dismissal of South Australia's pine plantations in 1957.

Mark Anderson learnt several European languages in order to read the forestry literature. Concerned about the lack of forestry text books with an ecological approach he wrote two (*The Natural Woodlands of Britain and Ireland* and *The Selection of Tree Species*) and translated Tamm's *Northern Coniferous Forest Soils* and Petrini's *Elements of Forest Economics* from Swedish; Knuchel's *Planning and Control in the Managed Forest* and Köstler's *Silviculture* from German; Larsen's *Genetics in Silviculture* from Danish; and Schaeffer, Gazin and d'Alverney's *Silver Fir Forests* from French. His many reviews, such as that on Cajander's book on Finnish forest types, also highlighted the ecological approach for English language readers. He was expert in the old Scots language and edited a book of *Proverbs in Scots* (1957). His monumental *A History of Forestry in Scotland* was published posthumously.

of the land was allied with the use of introduced species so that, by the 1960s, Sitka and Norway spruce covered a greater area than the native Scots pine.

Government grants and generous tax concessions stimulated a boom in converting the Scottish moors to plantations and supporting other types of forestry to the extent that the forest cover in Scotland had been raised from about four per cent in 1750 to over seventeen per cent today; for Britain as a whole it is about twelve per cent. Much of the increase created large blocks of single species, with little attention it seems being paid to the smaller scale needed to match species with sites on the intricate ecological basis that Anderson had developed.

Experience

The conversion of forests by replacing their species as in Central Europe, or by reshaping the land as in France and Scotland, or reforesting it as in the Karst region, became tools that foresters used widely, with various adaptations and mistakes to raise wood production elsewhere.

Foresters were attracted by the possibility of turning apparently worthless lands to good effect by the success of the plantations in Les Landes de Gascogne. They knew the story of stabilising the drifting coastal dunes and planting maritime pine and sought to emulate it without allowing sufficiently for different conditions or perhaps not understanding that the successful plantations at Les Landes were on the inland moors, well back from the coastal dunes. For example, a small plantation of maritime pine was started on coastal dunes in Western Australia in 1896 but it failed to grow. It was not until the 1920s that soil was analysed and foresters learnt how to select those areas where maritime pine could be grown successfully if the trees were fertilised.[15] In another example, a large experiment in both forestry and penal reform was started in 1913 in New South Wales.[16] It was a period in which rural work, far from the sinful cities, was seen as medically and morally reformative.[17] Coastal wasteland, unsuited to farming, was chosen and the work was done by prisoners who planted maritime pine and radiata pine. Although the trees grew well at first, the sand lacked zinc and other nutrients and the plantations failed. However, as a penal experiment it was repeated successfully in other areas.

Foresters were equally attracted to turning apparently worthless forests to good effect by converting them to conifers. There were two aspects to this. One was where the indigenous forests were naturally of little or no economic value; the other was where they had been so degraded by the *tire et aire* system or selective logging that the only practical way to restore their productivity was to clear fell and regenerate them with the existing species or replacement species. These became worldwide practices that accelerated from the 1950s under the push for economic development, as described in later chapters. However, the great extension of modern scientific forestry from its European roots was carried along in the expansion of European empires, as described in the next chapter.

Notes

1. Johann *et al.* 2004 gives a comprehensive overview.
2. These proportions are of the total land area and include areas that never carried forest, such as those above the tree line. See Smout and Watson, 1997; Tsouvalis and Watkins, 2000.
3. Mantel 1990.
4. Johann 2001.
5. The increases occurred even though the present share of forest land differs quite remarkably, from about 11% (Denmark) to more than 55% (Slovenia) of the total land.
6. Seeds were provided of peach, ash, maple, lime, swiss stone pine (arolla), apricot, chestnut, larch, acacia, elm and wild cherry. The range of species being grown gradually increased to include poplars, willows, black and cluster pines.
7. Reed 1954, Barrère *et al.* 1962.
8. Chambrelant 1887.
9. Temple 2011.
10. Barrère *et al.* 1962.
11. Schleicher 1917
12. See Oosthoek 2013 for an overall history.
13. Watts 2006 gives a short overview of the extensive research by Oliver Rackham and George Peterken and others.
14. Anderson 1932, 1950.
15. See Chapter 10.
16. Taylor 2008.
17. The culture of social reformation through rural work led to schemes that shipped British children to farm schools in Australia, Canada and Rhodesia (now Zimbabwe). A scheme for immigrant boys to establish plantations on the bleak coastal dunes of the south-west coast of Tasmania was proposed by the Conservator of Forests L.G. Irby in the 1920s to 'let the waste children of the Empire reclaim the waste lands of the Empire'; thankfully it came to naught.

7.

CONQUERING

European nations – Britain, The Netherlands, France, Germany, Spain, Portugal, Italy and Belgium – formed great empires by conquest or colonisation that spanned the world; the United States annexed some of Spain's to form its own; and Russia extended east, west, north and south to create its own vast realm. With these empires came their forests, their resources and their people. The conquerors brought their commodities, religion and the confidence that they could improve the world through modern science, but they also demanded resources in a hurry. The foresters' European vision of regulating use and managing the forests to ensure their perpetual productivity was tested by the greater complexities of imperial polities and populations, and by the greater diversities of the forests they encountered. How they advanced forest science to deal with them as best they could is the subject of this chapter.

The origins and spread of empires and their emerging interests in forest resources were highly uneven. Although empires had controlled forests to obtain timber and raise revenue from their early days, and forests had started to be conserved in Mauritius and the Seychelles from the middle of the eighteenth century, it was not until the early nineteenth century that the need for modern forestry started to take hold in major possessions.[1] Doctors and botanists in the great British and Dutch East India companies urged the need to control deforestation because they feared it would desiccate the climate and affect health and officials feared that the export trade in timber was depleting the forests on which it depended. Their fears centred on the teak forests of South India, Burma and Java. Teak timber, resistant to weather and termites and easily worked, had been highly-prized and internationally traded for at least two millennia. In the age of empire it was keenly demanded for ship's decks and superstructure, laboratory benches, windows, doors and garden seats and, from the 1850s, for the expanding Indian railway system. Trading companies secured leases over forests to extract the teak from areas nearest to rivers and ports, gradually moving inland as they exhausted the close resources. Their elephants hauling and cleverly stacking the logs became popular emblems of the trade, but safeguarding the forest was essential if the trade was to have a future.

From the mid-nineteenth century colonial governments started to organise forestry as a state function in its modern, scientific form and for that they needed trained foresters. They looked first for foresters from the German forestry schools and some from the French national school. For example, the Dutch recruited three German foresters to Java in 1849 and the British recruited a series of them, such as Dietrich Brandis in 1856, Berthold Ribbentrop and Wilhelm (later Sir William) Schlich for their Indian empire (now India, Pakistan, Bangladesh and Burma) where they created the exemplar of 'imperial forestry' in what became the world's largest forest service. The other empires followed suit with the Germans in Tanzania in 1898, the French in West Africa in the 1920s and the Belgians in the Congo in the 1930s, for example.[2]

British India

The foresters coming to British India and other empire possessions faced all the problems of measurement, silviculture and regulation that they knew from their European training but they had to adapt the accepted methods to different circumstances. It was far from a straightforward transfer of knowledge from the core nations of the scientific world to the colonies. They developed new methods at the same time that the foundations of forest science were still being elaborated and some of their methods influenced European forestry and spread to America and other lands of new settlement.

The foresters' major concern through much of the imperial world was how to manage forests in which a few commercially valuable species, such as teak or mahogany, grew among other species of no immediate commercial value. The start of all forestry knowledge, how to identify the trees growing in such forests, was an immediate and large problem. In Central Europe the botany and uses of the thirty or so species that were commercially important were well known; even with the species introduced for plantations there were only fifty or sixty that foresters had to know. In a tropical rainforest, there could be as many in a single hectare, many might not have been named scientifically and they were often hard to distinguish. Like many other countries, Britain sent students to France and Germany for forestry training but, as its imperial, and domestic, needs for foresters increased, it started its own English-language forestry schools, first in 1885 at the Indian Engineering College in England and then at the Universities of Oxford

Figure 7.1a. Landscape in North-West India (now Pakistan), at the foot of the Himalaya. A sal forest in the background.

Source: From a water-colour painting by Lady Katinka Brandis, taken from a reproduction in Schimper 1903, *Plant-geography on a Physiological Basis.*

Figure 7.1b. The temperate rainforest of the Eastern Himalaya (montane region, about 2,400m) near Darjeeling.

Source: From a water-colour painting by Lady Katinka Brandis, taken from a reproduction in Schimper 1903, *Plant-geography on a Physiological Basis.*

and Edinburgh. The imperial orientation of these courses was set by their professors: Schlich, a former Inspector-General in India and E. P. Stebbing, a former Conservator and forest entomologist there. They included tropical botany in their syllabuses and Schlich produced a five-volume *Manual of Forestry* for his students that became a standard textbook throughout the English-speaking world.

In the nineteenth and early twentieth centuries the duties of foresters and botanists often overlapped. Although the foresters took courses in botany, they needed the botanists of the colonial and imperial botanical gardens to identify species they were unsure of and to name those that were unknown to science. Trees were under-represented in the botanists' herbaria because the flowers, buds and fruit needed for identification were difficult to collect. To get the trees named, the foresters found that they had to be the collectors, while the botanists found that the trees became more systematically included in their herbaria. The botanists' training often included forestry and some offered lectures in forestry before forestry schools were established. Several colonial botanical gardens grew the tree seedlings for the foresters and some botanists in smaller colonies were responsible for the forests.

The size of the botanical identification task can be gauged from Brandis's 1906 list of 4,000 Indian trees, shrubs, bamboos and climbers, or from R.S. Troup's magisterial 1921 treatise on Indian silviculture that covers 800 indigenous and introduced trees. No other possession had received as much attention for so long and the task had to be taken up anew as forestry was extended. For example, when Charles Lane Poole was appointed as the first Forest Conservator for Sierra Leone in 1910 he found many trees had not been identified. He collected specimens which he sent to Kew where 21 of them became the 'type specimens' that identified their species as new to science.[3] In the 1920s he spent three years exploring the commercial potential of Papua and New Guinea's forests. Again he collected specimens, sending them this time to the Queensland Herbarium where 45 were identified as new to science.[4]

The imperial foresters had to deal not only with different species, but also with *types* of forest unknown in Europe – from dense rainforests to sparse arid woodlands, from coastal mangroves to lichen-festooned moss forests high on tropical mountains. They also found familiar conifer forests, albeit with different species, in the Himalaya and tracts of gregarious hardwoods, like sal forests, that seemed similar to European types. It was not until 1898 when Andreas Schimper published his treatise, with its English translation

Figure 7.2. Forest assessment party in front of an ilimo tree in Papua, 1923. Note packs and box for herbarium specimens in the foreground.

Source: Photograph by E.R. Stanley, National Library of Australia MS3799/3/glass slide.

in 1903 as *Plant-geography upon a Physiological Basis,* that they had their first scientific classification of the world's forests.[5] Schimper based his classification on the principle that 'the differentiation of the earth's vegetation is thus controlled by three factors – heat, atmospheric precipitation (including winds), soil'.[6] He not only provided broad categories in the major climatic zones – rainforest, monsoon forest, savannah forest and thorn forest in the tropics – but also provided principles for classifying the regional and local forest 'formations' by their climate or soil. He gave the foresters the structure and terms that they needed to refine the broad categories for their regional types in terms of the few things they could measure and observe: rainfall, temperature, elevation, geology and soil. It was a long process that depended on the gradual accumulation and mapping of the basic data across each country, but gradually they could get a better understanding of the complex patterns in their forests.

The foresters had to control the forests to implement imperial governance as well as to regulate how they were used. It was far from simple. Economically, there were increasing demands for wood for railways and export and from colonial treasurers for revenue. Politically, there were different and changing principles of governance ranging from direct rule, where state forests could be declared, to indirect rule, where princes or chiefs had to be persuaded. Socially, many forests had long been used by local peasants as part of their subsistence economy; there were forest-dwelling people who subsisted by hunting and gathering living in the forests; there were shifting cultivators who cleared patches of forest and moved on every few years and sacred trees, groves and shrines abounded. On top of all this was the gradual increase in the population with its demands for fuel wood and land.

The foresters' first concern was to gain control. The process they developed in British India was applied in a broadly similar way elsewhere.[7] Their first step was to classify the forests into the most valuable ones that they should manage permanently as state forests, others that should be protected from clearing for agriculture and small areas to be set aside for village use. The next step was to 'settle' the traditional uses of the best forests to curtail and progressively eliminate them by force, compensation or accommodation by turning any rights into permissions. The last step was to 'demarcate' the best forests by surveying and marking their boundaries and have them legally declared as reserved state forests under forestry control. Once in control, the imperial foresters started to close access, remove people and impose licence fees on every use of the forests from cutting timber to gathering honey. The extent to which they could do so varied from place to place as pre-existing users frequently had to be allowed some limited rights or compensation.

Once their boundaries and revenues were assured, the foresters could start to assess the resources and control their exploitation. Controlling what was being cut in the forests was the most urgent task. The simplest way to conserve young trees was to ban the felling of trees below a decreed diameter. Such decrees had been issued in Java and Bengal at the end of the eighteenth century, but to set them in a system of scientific forest management required the quantitative assessment of whole forests.

The foresters encountered problems in assessing the different types of forest, but everywhere they had large tracts with few if any previous maps to guide them. They generally used the strip survey method (introduced in Chapter 1) to map the interior streams, ridges and other features in the forest while collecting the resource information. With so many species, the foresters had to learn how to pick out the main ones by the appearance of their canopy, leaves, bole, bark and wood. They recruited local people whose intimate knowledge could often do this consistently and the foresters could then relate their local names to the scientific ones. In many types of forest, the undergrowth had to be cut along the centre lines of the strips spaced regularly across the forest before the assessment crew could get through to inspect and measure the trees. Even the simplest measurement of diameter could be difficult on trees with great buttresses. Rather than measure every tree, the foresters found they could judge the diameter of most of them well enough to put them into broad classes. Height was generally impossible to measure in moist tropical forests with their dense canopies. Only when the trees were felled could their length be measured. It was important to know the age of the trees in order to work out how fast they were growing and calculate the productivity of the forest. Age was easy to determine in Himalayan mountain forests with cold winters, where the trees formed distinct annual rings as they did in Europe, and the age of plantations was obvious from their date of planting. Other seasonal changes could cause annual rings, as the monsoon-dry period cycle did for teak, but in many other species and places the foresters had to rely on the slow process of comparing successive measurements taken years apart.

The first simple model to regulate the exploitation of the teak forests, based on diameter assessments and age estimates, was developed by A.V. Munro in the princely state of Travancore in South India in the 1820s. He tallied the assessments into six diameter classes ranging from the big trees, which took one hundred or more years to grow to over eighty centimetres, to his youngest class which reached thirty to forty centimetres in about as many years. The diameter limit below which trees could not be felled had been set far too low at twenty centimetres, but Munro and his successors gradually increased it to eighty centimetres by 1870, enabling large trees to be grown again in the forests.

In spite of this and some other early examples, it was not until the 1850s that science-based forestry became established across British India under the leadership of a Scottish Governor-General, Lord Dalhousie, and a German university-trained, botanist-forester, Dietrich Brandis. Dalhousie, an energetic and efficient administrator, set out clear principles that licence conditions were to be upheld in spite of complaints by the timber compa-

Box 7.1. Sir Dietrich Brandis (1824–1907)[9]

Dietrich Brandis, the son of a famous professor of philosophy, was born in Bonn in 1824. He studied natural science and botany in Copenhagen, Göttingen and Bonn where he gained a Ph.D. in 1848. He moved to India to work as a botanist and in 1854 married Rachel Marshman, the daughter of an orientalist and missionary. His brother-in-law and dedicated Christian, General Henry Havelock who had fought in the Anglo-Burmese War (1823–1826), was consulted by the Governor-General, Lord Dalhousie when he needed someone to take charge of the forests in the recently-annexed province of Pegu in Lower Burma. Brandis was appointed Superintendent of Forests in 1857, and the next year, at the age of 37 he became responsible for all the forests of Burma. Although there was a remarkable resistance from the administration as well as the timber merchants, he introduced the first plan for to control logging, and he organised a central timber market in Rangoon.

Portrait of Dietrich Brandis

Forest Research Institute, Dehra Dun. Wikimedia Commons.

Because of his success in Burma, the Indian Central government appointed him as Inspector-General of all the state forests in British India in 1864. Together with Hugh Cleghorn, the first Conservator of the forests in Madras, he was responsible for the *Indian Forests Act* of 1865. He established the Indian Forest Service and research stations across the country. To staff it he recruited European trained officers, most notably Wilhelm Schlich and Berthold Ribbentrop from Germany, who one after the other became Inspector-General of the Indian Forest Service. He also organised the education of British foresters in Germany and France. His conviction of the value of good forestry education resulted in the first college for Indian foresters being established at Dehradun in 1878. It was upgraded in 1884 as the Imperial Forest College and in 1906 became the Forest Research Institute and Forest College that still exists today.

His first wife died 1863 in Simla. In 1867 he married Katinka Hassel from Bonn. She accompanied him on his journeys in India and assisted his work with her excellent paintings of flowers and forest plants. She also provided them for an English edition of Schimper's book on forest ecology (Figure 7.1). They had four sons and three daughters, but three of their children died young.

He retired from the Indian Forest Service in 1883. Two years later the first British forestry course was established by Schlich at the Royal Indian Engineering College at Coopers Hill with Brandis as a visitor. He devoted the remainder of his life to his botanical studies, which he had managed to maintain in India. He was honoured with a British knighthood in 1887, an honorary doctorate from the University of Edinburgh in 1889 and an honorary professorship in Prussia in 1890. He managed to complete his great work, *Indian Trees* the year before he died in 1907.

nies, that a long-term view was to be taken for the 'proper maintenance' of the forests and that the teak forests were state property. In 1856 Dalhousie appointed Brandis to take charge of the major teak producing region at Pegu in Burma.

Brandis had the same type of information as Munro, but he collected it more systematically. He laid out his strip surveys on a regular grid, rather than on a convenient 'line of march' through the forest. He took detailed measurements on the felled trees so that he could prepare volume and yield tables. He devised a simple model in 1857 to regulate the yield on the same two simple factors as Munro: age and number. He found that it took 25 years for a teak tree at Pegu to grow to fifteen centimetres in diameter and a further sixty years for it to grow over sixty centimetres into the largest of his four classes. As it took thirty years for the second class of trees to grow into the largest class, he could safely fell one thirtieth of the largest trees each year, knowing that they would be replaced in time. As he and others gained experience with this rule, they made adjustments to allow for exceptionally large trees, trees retained to provide seed, variations in numbers across the forest and so forth. Brandis was well aware of the theoretical inadequacies and called for permanent sample plots to be installed so that future calculations could be based on the forest's volume increment. Nevertheless, the rule's simplicity led to it being widely adopted throughout the tropical world.

Although the rules developed by Munro and Brandis provided a quantitative basis for regulating the yield from the forests, they and the other imperial foresters had to find ways to regenerate them after trees had been felled. First, they had to learn the 'silvical characteristics' of the multitude of species they encountered: what soils and climate suited each best, how much and how often each produced seed, how well it germinated, how the roots developed, whether the young trees needed light or shade, whether they produced coppice shoots when cut and so forth. Then they had to devise a 'silvicultural system' that would meld the characteristics of the trees with ways they could cut the forests. They had the sheaf of European systems to draw on, but these were different circumstances.

Individual foresters observed how the different types of forest responded after trees had been felled and they tried sowing seeds, transplanting forest seedlings and freeing young trees from vines and weeds. Gradually, they accumulated sufficient practical experience so that in 1870, the chief foresters in the different Provinces could hold a conference, and 1875 start their own journal, *The Indian Forester*. However, absorbed in the day-to-day management of the forests, they could not advance their scientific base beyond observation and simple trials until an Imperial Forest Research Institute was established at Dehra Dun in 1906. The Institute was regally housed – jointly with an older college for training Indian subordinate foresters – and staffed with specialist scientific officers to conduct research in silviculture, zoology, botany, chemistry and economics. An Imperial Superintendent of Working Plans was included to oversee the regulation of the forests throughout British India, and the Imperial Silviculturist was supported by silviculturists appointed in several provinces.

Silviculture

Many of the silvicultural problems arose because the timber industry demanded only the finest of the many species in the forest. For example, it took only the teak trees for the export trade, yet even in the best teak forests they only ever amounted to one third of the number and in the poorest only one sixteenth. Such highly selective logging made it hard to regenerate the forests naturally with light-demanding species, such as teak or eucalypts, which needed bigger openings in the canopy to let the young trees grow. Moreover, vigorous competing species such as climbers and bamboo in many of the teak forests might capture the openings first. Brandis introduced the *taungya* method into imperial forestry as a way of overcoming these problems by allowing peasants to clear patches and cultivate their crops for a few years before planted or regenerated trees took them over.

The foresters and research officers tried to develop silvicultural systems aimed not only at regenerating the forests after logging, but also at 'enriching' them by increasing the proportion of the commercially valuable species.[8] Just as every generation of foresters had before them, they needed to find the best conditions to germinate seed and grow seedlings. This was the easiest part, as the silviculturists could make small, designed experiments in the forest centres and have results to report in a year or two. They also photographed the seedlings to help identify their species in the forest.

They took three routes with their developments. The first aimed for entirely natural regeneration and tried ways of enlarging the gaps opened by logging, burning the competing vegetation and preparing the soil for the seed, naturally falling or sown. The second aimed to 'enrich' the forest by planting the most favoured species in gaps or lines cut through the forests. Both these routes needed to have the weeds and climbers cut away from the young trees several times, which was difficult to do and verify. The third

Box 7.2 *Taungya*

Foresters everywhere were completely opposed to the shifting cultivation of 'slash and burn' agriculture in the tropical forests whereby peasants cleared small patches, burnt the felled trees, grew their crops for a few years until the soil was exhausted and moved on. The abandoned patches were left fallow for three or four decades, during which a 'secondary' forest of young trees and weeds naturally recaptured them. Importantly for the foresters, the species they most desired did not always return. They saw it as destruction, but their ability to evict the peasants and deprive them of their means of subsistence was sometimes limited. Instead they sought to put it to good effect.

Once state forests had been demarcated and the foresters had control, they could use the *taungya* system. For this, they only allowed the peasants to take up patches in logged over forests for three or four years, on the condition that they sowed tree seed among their crops. The peasants could grow their crops until the young trees started to close their canopy and shade them out, leaving the land to grow only timber. Teak was the foresters' favourite, but they had to return two or three times to clear away weeds and climbers. It was an ingenious system for converting the natural forests into a pattern of mini-plantations of the desirable species, at very little cost to the forestry departments; the peasants did the hard work of clearing the land.

As an imperial silviculture it is credited to Brandis who in 1856 inspected a teak plantation that had been established in this way by a Karen from northern Burma. He adopted the system, which spread to other parts of British India and South Africa by 1887. The Dutch introduced it successfully as *tumpang sari* to Java in the 1870s and 1880s; it was used in Thailand from 1906, and introduced into Nigeria in the 1930s.[10]

The *taungya* system had several older roots in China.[11] One in the seventeenth century described establishing stands of Cunninghamia or 'Chinese fir' by using food crops to shelter the young tree seedlings. The Chinese and imperial forms resulted in tree plantations without food crops and are distinct from the numerous agro-forestry systems that co-produce wood and food. Its imperial form in the nineteenth and twentieth centuries successfully married the foresters' need to regenerate and enrich the forests with the imperial need for social control.

aimed to replace the existing forest with a favoured species, either indigenous or introduced. Plantations were costly to establish, had the same problem with weed competition and for some species were more liable to disease and insect attack, but at least they could be supervised easily and the results were clearly apparent. Scientifically, most of the development of these systems probably first occurred at the level of local trials and observations – 'observational experiments'. Formally designed experiments were difficult to conduct in the highly variable tropical forests and most remained at the level of long-term, measured and reported trials, rather than as replicated experiments that could be analysed statistically.

Imperial forestry

The progress of forestry was interrupted by World War One, after which Germany's colonies were distributed among the other empires. By the 1920s virtually all except the smallest colonies had started to put modern scientific forestry in place. Although forestry was a colonial, not an imperial government function, it took on the same features of state controlled and professionally managed forests everywhere and was designed to serve both imperial and colonial interests. However, the extent to which a scientific and quantitative base was established varied widely.

British India was the most advanced, with assessments, permanent sample plots, volume and yield tables, and a well-established Forest Research Institute. Malaya followed the Indian experience by setting up its own Research Institute in 1925, installing growth plots and successfully researching ways to regenerate its dipterocarp forests with shelterwood systems. Foresters were appointed as specialist silviculturalists in several other colonies. For example, two were appointed in Nigeria in 1920 where they experimented with different ways of regenerating the forests. The scientific progress of tropical forestry was not limited to these dedicated institutes and specialists; indeed it was inherent in the ethos of all colonial forestry departments, which developed systems to suit their particular forests. For example, the French colonial foresters developed several different systems in their African colonies.

The era of imperial forestry had carried and adapted modern scientific forestry into the tropical world and had started to understand some of its complexities. In many cases, imperial power had controlled native populations more successfully than the foresters had been able to rein in the economic and political power of timber companies where, especially in Africa, they had been granted extensive, long-term concessions. However, the advance of imperial forestry was restricted by the economic depression of the 1930s and interrupted by World War II. When the empires gradually gave way to the post-colonial world of the 1950s and 1960s, forestry had to adapt to the new realities. How it did so is discussed in the next section of this book.

Notes

1. Grove 1995.
2. Sanseri 2009, Haig *et al.* 1957, vol. 1.
3. Dargavel 2008.
4. Dargavel 2006.
5. Schimper 1898 in German appeared in English in 1903 and in a revised German edition in 1935. The 1903 edition was translated into English by W.R. Fisher, the Assistant Professor of Forestry in the Royal Indian Engineering College at Cooper's Hill in England (the precursor to the forestry school at the University of Oxford). Schimper was the first to use the term 'rain forests', but it was not until 1957 that a detailed classification of them was prepared by Richards.
6. Schimper 1903.
7. The literature on the development of Indian forestry is extensive; see for example Barton 2002 and Ravi 2006. Buchy 1996 gives a detailed account of how it applied in South India.
8. Haig *et al.* 1957–1958. Dawkins and Philip 1998.
9. Information drawn from Rawat 1993; Prain 2004.
10. Boomgaard 1988; Dawkins and Philip 1998.
11. Menzies 1988.

DEVELOPMENT

The foresters' sylvan vision of sustaining the wood yields from the natural forests was superseded by the urgency of demands. From 1950 to 1970, the world's population increased by 45 per cent, it used 57 per cent more industrial wood and showed every sign of increasing. But there was an optimistic mood to the post-war era: the ravages of war were being repaired, the cities rebuilt, displaced people given homes and the economy was expanding. Science and technology were spectacular: nuclear energy, used so terribly in Japan, promised unlimited power, a spacecraft orbited the earth, a man walked on the moon and computers became widely used, although it was not until the 1970s that they became ubiquitous. With science, technology, peace and good intent an even larger world could be developed to be a better place, or so many hoped.

World polity changed too, as European empires gave way, peacefully or bloodily, to the nations formed from their colonies, while the Soviet Union challenged America's might. Hope found its expression in the United Nations, created in 1945. It saw promoting economic and social development, particularly in developing countries, as a means of maintaining peace. The Food and Agriculture Organisation (FAO) was responsible for aiding development in the forest sector and became a major focus for forestry throughout much of the world. It continued the World Forestry Congresses, first started in 1926, disseminated information, funded projects and sent experts to advise developing countries on how to enact them. Development aid and expert advice was also provided bi-laterally by several developed countries. The source of this expert advice lay partly in European forest science, partly in what had been developed in the former empires and in lands like Australia and Canada and increasingly in American science and technology. Although forest research was almost entirely a public activity in state research centres and universities, an emerging feature was the involvement of large American companies, either directly or in cooperative associations in research.

Forest science flourished. The foresters realised that their vision of reaching the potential of the natural forests still had to be extended into much of the tropical world, but that it could be bolder. They already had a history of raising productivity by introducing new species and converting land into productive forests, but the two World Wars and the Great Depression between them had curtailed their hopes. Now with more money and new research methods and technologies they could advance their science and apply it. Although they advanced on many fronts, it was the bold expansion of industrial plantations that epitomises the period and that is the focus of the next three chapters.

The foresters took three new routes to raising productivity. One was to grow better trees. They already had experience in introducing new species and by the 1950s knew which did best in most places, but they could improve their growth by intensive breeding (Chapter 8). Another was to make the trees grow faster by fertilising them. They had conducted some experiments to get the trees to grow at all on some of the poor Scottish moors, but with sufficient finance they could use fertiliser to increase growth as a normal practice (Chapter 9). Closely allied to this was to use chemicals to control weeds. In these two routes, forestry was belatedly following the path of agricultural development. The foresters' third route was to improve the way they planned the management of the forests was and this was enabled by computerisation (Chapter 10).

8.

BREEDING

The intense debates about evolution and diversity of the natural world and its peoples were coincident with the foundations of modern forest science, described in earlier chapters. Well-known markers in the advance of science occurred when Charles Darwin's and Alfred Wallace's revolutionary ideas on evolution were presented in a public lecture to the Linnean Society in 1858; when Darwin's *The Origin of Species* was published the next year; and when Gregor Mendel gave his paper on the genetic laws of inheritance in 1865, although it was not until his paper was rediscovered at the start of the twentieth century that it became widely known.[1] An apparent rift emerged between Darwin's theory that species gradually adapted to their environments – through mutation and the 'survival of the fittest' – and Mendel's theory of genetic determination. It was not until the 1920s and 1930s that Ronald Fisher and others were able to bring the ideas together in a synthesis founded on statistical analysis.[2]

The foresters' investigations were practically, rather than theoretically, driven.[3] They sought ways to manage diversity through 'nurture' in their silvicultural systems, or turn 'nature' to good genetic effect by introducing species from other lands (Chapter 5) or by breeding better trees. They could alter the size, form and appearance – or 'phenotype' – of the trees through silviculture, but had negligible or only very long-delayed effects on the genetic composition – or 'genotype' – when they relied on seed falling naturally from the mature trees. It was only by selecting which trees to breed from and which seedlings to plant that they could influence genetic variation significantly in any reasonable time. Their investigations evolved through stages from simply collecting seed from particular individual trees, to collecting seed grown in particular places, to establishing 'seed orchards' to produce seed from superior parents and to the most recent investigations into biotechnology. There are some recent breeding programmes for non-industrial objectives; however forest tree breeding developed as a plantation activity, most strongly for large industrial plantations, often with introduced species.[4]

Chapter 8

Individual trees

It seems so obvious to choose a good tree if one is collecting seed. Certainly Evelyn in the seventeenth century thought that the differences in the growth of young trees were due to the 'variety and quality of the seed' and therefore that seed should be collected from trees 'such as are found most solid and fair'.[5] He considered that a 'Microscopical examen to interpret their most secret Schematismes' would be an 'over nicity' [*sic*] for plantations, as indeed it was in the seventeenth century; only in the twenty-first century have the secrets of DNA started to be revealed for trees. However, Evelyn was not quite clear and eighteenth-century writers often confounded the characteristics of the parent tree with the size of the seed, the type of soil and other factors.

The first really clear statement of the difference between tree nature and nurture – between genotype and phenotype – was made by the multi-talented Frenchman, hailed as the 'father of French silviculture', Henri-Louise Duhamel du Monceau in 1760.[6] He wrote that the seeds of stunted, ugly trees in the forest would produce ugly offspring if that was their true nature, but if their irregular form was due to some accident, then their seeds could produce beautiful trees. He used the familiar avenues of elm trees growing along French roads to explain the matter.[7]

Duhamel had collaborated uneasily in forest experiments with the great encyclopaedist, Georges-Louis Leclerc, Conte de Buffon early in their careers, but developed his understanding independently of Buffon.[8] Duhamel and Buffon were great collectors; Duhamel accumulated an arboretum and plantations of nearly 700 species on his estates, and Buffon was Director of the Royal Botanic Garden in Paris. Each was a member of both the French Academy and the British Royal Society. Not only were there common interests in trying to understand the differences and similarities found in the natural world in general, but there were common interests in trees.[9] Duhamel, like Evelyn was concerned about the supply of timber for his country's navy and, as the eighteenth century progressed, landowners in these and other countries established plantations and planted rare and interesting trees to grace their estates. Several large and apparently prosperous firms of seed merchants, nurserymen and landscape gardeners arose to supply them. It was from this tradition that the first scientific advances in breeding trees originated.

Provenances

Nurseries to restore forests were called for by the famous mining officer Hans Carl von Carlowitz in Saxony (now Germany) in 1713. Shortly after, in 1727 Carl, Count of Hessen established the largest nursery of the period at Kassel. As interest in introduced species increased in the second half of the eighteenth century, two aristocrats started importing tree seeds from North America.[10] By 1783 there were 68 different species on sale, and by 1790 there were 253. August von Burgsdorf established one of the most famous nurseries near Berlin. He collected and imported seeds from America and sold trees from 1786.

In France Philippe André de Vilmorin diversified the well-established plant and seed supply firm of Vilmorin-Andrieux into supplying forest trees at the start of the nineteenth century. He started by establishing a small arboretum outside Paris in 1815 and expanded by establishing a larger one at Des Barres in Central France (now the state *Arboretum National des Barres*) in 1821. Although he collected many individual species of conifers and hardwoods, like other horticulturalists of the time, he was intrigued by the significant differences in growth and form between Scots pine trees growing in different parts of Europe.[11] Although the differences had been known for a century, opinions varied widely: some people denied any natural distinction and attributed the differences to variations in soil, climate or external factors; some thought that there were two or more species; while others thought that there were different varieties or races that bred true in successive generations.[12] Vilmorin decided to test the matter scientifically by obtaining thirty lots of Scots pine seed from different sources in France, Russia, Scotland, Germany and Switzerland. After growing them in small plots at Des Barres for up to thirty years he was able to classify the lots into five groups that gradually changed from the slender pyramidal shape of the Russian pines to the spreading shape of those from southern France.[13] His nurture of each lot had been the same, but their nature had differed according to their origin. This genetic variation between seed lots of the same species from different places became known technically as their 'provenances'.

This understanding gave foresters both warnings and opportunities. They had to be careful where they got their seed, because the trees they grew might not be all they expected. For example, in 1855 Swedish foresters warned landholders against buying German seed because the trees were less resistant to fungal disease and in 1888 the state imposed an import duty to limit the practice. New Zealand's experience provided the most salutary lesson when it bought seed of ponderosa pine from an inland Canadian source in the 1930s, rather than from California or Oregon, only to find that its several thousand hectares of plantations were of a variety that grew so poorly that they had to be replaced.[14]

The prospect of distinguishing provenances raised the foresters' hopes that they might be able to introduce a provenance that was superior to their local one. But this had to be tested. Investigations such as those by Adolf Cieslar in Austria had shown that provenances varied by altitude due to climatic differences and hence that they had evolved genetically in response to them (Figure 8.1). Because there were different environments, the trials like those at Des Barres had to be repeated on different sites, and to give scientifically valid results they had to be statistically designed and analysed. The provenance trials they established in Europe, North America, South Africa, Australia, New Zealand and several other countries were not only large, expensive experiments, they also required international collaboration to select and distribute the seed and in some cases to conduct and analyse the trials themselves. This started a necessary culture of collaboration that persisted through the subsequent stages of developing tree breeding.

The first international trial was planned in 1900 at the International Union of Forest Research Organisations' (IUFRO) Congress to evaluate provenances of spruce, Scots pine, larch and oak collected from various altitudes and environments across Europe and Siberia.[15] It started in 1907 with a trial of thirteen provenances of Scots pine and was repeated in 1938 with 55 provenances of Scots pine and 36 provenances of spruce.[16] Several international provenance trials followed with the largest being stimulated in the 1960s by the need to find strains of spruce that best resisted air pollution. It compared 1,100 provenances planted in twenty sites in thirteen countries.

Breeding research

Although choosing the best species, provenance or tree from the diversity of what nature provided in the wild were clearly sensible measures, they gave foresters little control over the genetic composition of their forests. Forest trees had not been domesticated as agricultural crops or animals had been over the centuries; their parents had not been chosen and mated, generation after generation. Clearly it was difficult to do so with forest trees whose flowers are difficult to reach, yet, to control breeding, the foresters had to control pollination and they had to be patient as it takes years before trees are mature enough to produce seed.

Figure 8.1. Alpine forest trial plots near Aussee (Styria, Austria) at 1,400m above sea level, about 1900.

Source: Forstliche Bundesversuchsanstalt in Wien [Federal Forest Research Centre Vienna] 1974. *The History of the Federal Forest Research Station and its Institutes.*

The first organised programme to control pollination in forest trees was initiated by an American timber company owner, James Garfield Eddy in 1925. He knew fruit and nut trees were bred that way and he wanted to apply the techniques to pines and other conifers.[17] He personally funded a research station in California until 1935 when it became the U.S. Forest Service's Institute of Forest Genetics. Its first scientists were recruited from the arboriculture of fruit trees, genetics and plant physiology, rather than from forestry. Their initial objective was to breed hybrid species of pines, as had been done with American black walnut trees and as had occurred naturally with larch. To this end, they established an extensive arboretum and developed techniques for pollinating mature trees in the forest. These were laborious, as pollen-proof bags had to be put over unripe female flowers and pollen collected from male flowers and injected into the bags later. Their results were disappointing because the hybrids they bred from the species they tried were no better than their parents.

Breeding programmes started in other centres during the 1920s and 1930s, most notably in the Southern USA and Scandinavia. In Denmark, Carl Syrach-Larsen produced hybrids of larch and other species and took the important step of using grafts and rooted cuttings – 'vegetative reproduction' – to create clones of the parent trees. Although commonly done with fruit trees, it was a novelty with forest trees.[18] Rows of different clones planted alongside each other enabled him to display the differences between parents. However, it was the *combination* of grafting with controlled pollination that was the critical step. He showed that cuttings – 'scions' – taken from the upper crown of mature parent trees and grafted onto young 'stock' trees in an orchard, flowered readily and produced seeds five or six years later. As he did not have to wait twenty or thirty years to grow a mature tree from seed, it shortened the time needed to produce successive generations. It also made it much easier and cheaper to pollinate the small trees grown in an orchard. It was the key that would turn research experiments into the orchards that would eventually produce seed for commercial plantations. Before that could happen, the idea had to be proved.

Research was continued actively by three research groups in neutral Sweden during World War Two and resumed elsewhere afterwards. In spite of the reluctance of many foresters to consider more than silvicultural means of improvement, tree breeding became more widely known after its specialist journal, *Silvae Genetica* commenced in 1951. Scandinavia and the United States continued as the major research countries, but were joined by countries such as Australia and New Zealand as they expanded their industrial plantations.

Generations of orchards

The Swedish research groups had clarified a protocol of the stages of work needed to produce seed at an operational scale. First the best, or 'plus' trees had to be found in the forests. Then cuttings from them were grafted on to young stock to create a 'seed orchard' containing copies of each of the 15 to 100 parents. Once the orchard trees had started to flower, each parent was crossed separately with others by controlled pollination. The resulting seeds were raised and planted out in 'progeny trials' where their performance could be measured and the usefulness of each parent evaluated. Any parent whose progeny proved to be too poor was then eliminated from the orchard.

Box 8.1. Carl Syrach-Larsen (1898–1979)[19]

Carl Syrach-Larsen explaining the idea of natural clones with anemones in a beech forest.

Photograph courtesy of Greta Olsen.

Carl Syrach-Larsen came from a family deeply involved with trees. His father, Georg, was a head gardener of the Forest Botanical Garden, part of the Danish Royal Botanic Gardens. He was born in the head gardener's house and spent his boyhood among the garden's trees. When he finished his schooling, he attended the Royal Veterinary and Agricultural College, now part of the University of Copenhagen, which also administered the Gardens.

Syrach-Larsen graduated with a forestry degree in 1923 when he was 25, and continued at the University and Gardens as a researcher. He first researched the geographic distribution of the larches which led him into the genetic questions of how species varied. He was able to spend a year, 1928–29 in the British Royal Botanic Gardens at Kew. His study of its arboretum, and of the Douglas fir trees that had been grown in Scotland from the seed collected by David Douglas, turned his mind to questions of seed orchards and how to breed trees.

Even though Europe fell into the depths of the Great Depression on his return, funds from private foundations enabled him to start his research into crossing trees by artificial pollination. He also developed techniques of vegetative propagation. There was limited room in the Gardens and a new site was needed for his many experiments. In 1936 a new arboretum area was opened at Hørsholm; Syrach-Larsen was appointed its Director and the next year he was awarded a doctorate. He was able to continue with some research work even during World War Two.

Syrach-Larsen, with twenty years of research and publications behind him, became a leading figure internationally as interest in tree breeding grew rapidly after the war. He visited North America in 1946, New Zealand in 1949 and led an IUFRO research group. It was Mark Anderson's English translation of his book, *Genetics in Silviculture*, in 1956 that made the ideas and techniques of breeding trees readily available to foresters world-wide.

He continued to travel widely in Asia, Africa and Europe, as well as North America and was recognised for his pioneering work with professional and academic awards.

Meanwhile, the flowers in the rest of the orchard pollinated naturally to provide seed for the regular plantations. Even though there was always the risk of rogue pollen blowing into the orchard, the genetic quality of its seed was markedly higher than from the wild populations and was progressively improved as the poorer parents were removed.

The Swedish protocol was adopted worldwide as more breeding programmes were started, but it was in the Southern United States where it was applied most vigorously and on the largest scale. The forest industry was expanding rapidly in the region from the 1950s as fifty large companies built pulp and paper mills there. They needed wood, bought forests and old fields, and planted 400,000 hectares a year of pines for at least thirty years, with similar rates of planting and re-planting thereafter.[20] They had to make their plantations as productive as possible and the success of the Swedish work provided an example of what might be done. Although they were fiercely competitive for resources and markets, they were willing to join cooperative breeding programmes centred on universities.[21] The first in 1951 focused on slash pine. It was headed by Bruce Zobel at Texas A&M University and was supported initially by eight companies. The second in 1953 also focused on slash pine was based in the University of Florida. The third in 1956 with about a dozen companies focused on loblolly pine and was also headed by Zobel when he moved to North Carolina State University. These three university-industry cooperative schemes developed as very large programs, with each selecting about 3,000 plus trees on their company lands to provide them with an initial source of parent material to exchange between members. They designed separate orchards to suit climatic differences across the region.

As these and other breeding programmes progressed, six considerable difficulties were realised. First, it was simply not feasible to test all the possible combinations of parents in progeny trials: for example, with 100 parents, it would have needed 10,000 crosses to be made, raised, planted in small plots and measured. Many other designs were tried from pollination in the open or from a mixed batch of chosen males, to testing only within groups of parents. Common to them all was the need for rigorous statistical design and analysis. The second difficulty was trying to understand which characteristics were determined most strongly by their parents – their 'heritability' – and which by their environment. This also had to be estimated from the statistical analysis of the progeny trials. For example, Australian work with radiata pine showed that wood quality was highly heritable. Growth rate, volume production, tree form and wood quality were the characteristics of great interest, with disease or cold resistance being important in some cases. The third perplexing difficulty for tree breeders was an economic rather than a scientific one: what was the relative value of each characteristic? For example they could breed slash pine for resistance to rust disease, but the most resistant trees were not the fastest growing; conversely the fastest growing were not the most resistant; and there were many intermediate positions. The difficulty of deciding which trees to include in the orchards was dealt with in the North Carolina programme as each company developed its own orchards to balance the risk of disease with volume gain in its own way.

The fourth and scientifically difficult matter was deciding how many parents should be included to avoid the risk of in-breeding. If there were too few or if close relatives mated, their progeny might perform poorly. Such 'breeding depression' was found to quite noticeable in pines, a serious matter. Rather than the 15 to 100 parents thought sufficient in the 1950s, by the 1990s it was thought that at least 400 were thought necessary for the long term.[22]

It was the long-term issues of designing subsequent generations of seed orchards that raised the fifth and most scientifically difficult problems for tree breeders. As the progeny from the first generation of orchards established in the 1950s became old enough for reliable evaluation towards the end of the 1970s, it became possible not merely to remove poor trees from the existing orchards, but to make a second generation of orchards using selections from the best trees. And at the start of the twenty-first century the process was repeated to create a third generation of orchards and the real gains from half a century of tree breeding became apparent: the volume growth of slash pine had been increased by thirty per cent and its resistance to rust greatly improved.[23]

The sixth difficulty was how to balance gains that could be made in the most desirable characteristics with the need to retain some of the inherent genetic diversity in case it should be needed in the future. Forecast changes to climate, possible changes to the qualities demanded and an ethical concern for nature's variety made it important to conserve the genetic base of each species. The simple Swedish protocol had to be elaborated into various 'advanced breeding strategies' in order to deal with all these difficulties. Some breeders went back to the forests and selected more plus trees to widen the genetic base of their programmes, others revised their selections to give more emphasis to characteristics realised to be more important or that could be raised once growth rate had been improved. What was common to all these programmes was the high level of education in genetics, statistics and forestry needed to design and manage them. Forest genetics had become a specialist discipline that was taught, mostly at the post-graduate level, in the forestry schools of a few universities.

Propagation

Trees for the plantations of most species were propagated from seed, genetically improved when the orchards could provide it. Only poplars, willows, Japanese cedar, and Cunninghamia were usually propagated from rooted cuttings and they were the exceptions until the 1950s, when tree breeders realised that it might be possible to propagate other trees that way. If they could do so on a large scale, they could make whole plantations of genetically identical trees. These clones would be identical in size, shape, wood quality and whatever characteristics they had been bred for, save only for unavoidable variations in the soil. Their uniformity would make them ideal feedstock for the large mills and a boon for mechanised logging.

The story of their development can be seen with the eucalypts, the second most important plantation genus after the pines.[24] Although Australia is the home of over 500 species of eucalypts – there are only two endemic species outside it – their plantation importance is greater elsewhere. Brazil, India and China have the largest areas, with Spain, Portugal and South Africa having large areas and many other countries being engaged in planting programmes. It was only after a few Australian foresters had seen the Brazilian plantations in Brazil in 1961 that they became interested in eucalypts as a plantation species and it was not until the 1990s that they planted them on an extensive scale.[25]

Many of the plantations outside Australia were started with seed from only a few trees in botanic gardens. In consequence they had a narrow genetic base. For example, the blue gum planted widely in southern Africa for eighty years originated from only nine trees and the forest red gum planted in West Africa originated from a single tree in Madagascar. Some of the plantations suffered from the problems of in-breeding depression when subsequent crops were grown from the seed of previous ones. The problems were exacerbated where hybrids had arisen between similar eucalypt species. Before breeding programmes could be started it was essential to find the provenances in Australia best suited to their new sites. However it was not until the late 1960s, a century after Vilmorin, that Australian foresters started to study how eucalypts varied across their continent.[26] Provenance trials followed in many countries, co-ordinated internationally in the 1980s by a special IUFRO Working Group.

The possibility of cloning eucalypts from rooted cuttings had opened up in 1948 when two Australian foresters, J.M. Fielding and Lindsay Pryor, produced them in experimental glasshouses. About the same time a forester in Morocco discovered some that had occurred by chance. Trials followed in Australia, North Africa, Papua New Guinea, the Congo and Brazil, but it was not until the 1970s that the critical importance of age was realised. Unless the cuttings were juvenile, they would neither root properly nor grow normally. Once it was realised, they developed techniques to produce cuttings *en masse* by stimulating rooting with hormones and by providing the best temperature, humidity and light. Special 'clone banks' were set up where supplies of very juvenile cuttings could be harvested. A common way of doing so was to set out lines of parent plants that sent up coppice shoots when they were cut. These could be repeatedly clipped to provide the cuttings. Every few years these 'hedges' had to be re-established to keep them physiologically juvenile. The rooted cuttings were grown on into seedlings that could be planted out normally in industrial plantations.

The geneticists found that they could also propagate tissues taken from particular parts of eucalypts – the growing tips of shoots, germinating seeds, or even small pieces of leaf – by culturing them in a sterile medium under laboratory-type conditions. By the 1990s Chinese foresters were able to establish plantations with trees started in this way, and the method was being taken up in other countries.

Although the forest geneticists had found out how to produce plants for clonal plantations, they also had to tackle the thorny problem of deciding which and how many of the parent clones to include. If they chose too few, there were higher risks; some clonal plantations of poplars had proved susceptible to new strains of leaf rust disease, for example. However, if they chose too many, they might not fully realise the gains from breeding. Theoretical studies suggested that somewhere between seven and fifty clones should be used.[27] Planting them in a chequerboard block pattern was also thought to reduce risk. These ideas became the normal protocol for plantations established by clonal propagation.

Gains and conservation

By the twenty-first century breeding forest trees had become more diversified, more advanced scientifically and a normal part of industrial plantation forestry. It had been concentrated most intensively on about sixty species of acacias, eucalypts, firs, larches, pines, poplars and teak used for wood production, but other species and non-industrial uses had also received some attention.[28] It was directed mainly to improving growth rates, tree form and wood quality, but some programmes emphasised resistance to disease or air pollution or tolerance to salty soils. It was furthest advanced with pines in the southern USA, South Africa, Australia and New Zealand and with eucalypts in Brazil, but many other countries had programmes at earlier stages of development. Hybrid species of some pines and eucalypts had proved particularly successful. Scientifically, it was swept up in the genetic revolution that followed the discovery of DNA and the development of abilities to manipulate the genetic structure of animals and plants.

The gains achieved in practice had increased with time. The growth rate of the early Brazilian plantations was raised from an average of four cubic metres per hectare per year to over forty cubic meters per hectare per year over the course of the twentieth century. In the southern USA, the growth rate of slash pine was raised by nine to thirty per cent as the breeding programmes became more advanced.[29] This level of improvement probably applies to radiata and other pine species in major plantations.[30] Economically, the accompanying gains in straightness, branch size and uniformity were also significant. At the enterprise scale the gains were realised progressively as each mature stand was replaced with genetically better young trees and as these in turn were replaced with still more improved ones. Although breeding programmes could only be afforded by large companies or state forest services, their nurseries commonly sold improved plants to smaller growers.

A further problem for the foresters was how best to conserve the genetic diversity of each species. As the spectre of climate change became more apparent, they realised that they might need to find genes suited to new circumstances and they recognised the moral claim to preserve all of nature's variety. Compared to agricultural species with few if any areas remaining in the wild, theirs was a minor problem; selective breeding had not been practised nearly as long, and most species were well represented in conservation reserves. However, there were a few trees that needed special measures, notably radiata pine that had been extensively planted in New Zealand, Australia and Chile since the 1920s. Its natural habitat in five small widely separate areas along the Californian coast and off-shore islands had been greatly reduced. In 1978 new collections were made across the whole range to enhance the existing breeding programme and conserve the species.[31] However, conservation in the long term depends on new stands grown from these collections being perpetuated. Similarly, collections were made of a tropical eucalypt – rainbow gum or deglupta – in danger of extinction in the Philippines and some provenances of these were planted at three sites in Papua New Guinea.

The gains achieved by breeding forest trees for industrial plantations were accompanied by gains being achieved by fertilisation and chemical weed control, as described in the next chapter.

Notes

1. Kohn 2004 provides a history of evolutionary thought from an English perspective.
2. Edwards 2001.
3. For histories of forest tree breeding, see Syrach-Larsen 1956, ch. 2 'Retrospect'; the detailed ecological paper by Langlet 1971; the overviews by Burley 2004 and by Burdon and Libby 2006.
4. Kanowski 2001.
5. Evelyn [1664] 1995, p. 201.
6. Bonnaire 2000.
7. Duhamel, pp. 86–87. Syrach-Larsen, p. 16 provided the reference to Duhamel's book.
8. Fellows and Milliken, 1972, pp. 51–2.
9. Hartley 2010.
10. von Veltheim from 1750 and von Münchhausen 1753.
11. Vilmorin 1864.
12. Valmorin 1864, p. 32. noted an observation in the Earl of Haddington's 1761 *Treatise on Forest Trees* that supported the last opinion.
13. Vilmorin called the Russian provenances 'Pin de Riga'. Riga, now in Latvia, was a major port for timber exports.
14. Poole 1961, p. 46.
15. International Union of Forest Research Organisations 1992, p. 4.
16. König n.d.
17. Stone, 1968; Burdon and Libby 2000.
18. Syrach-Larsen 1956. Poplars were the exception as they are commonly raised vegetatively from cuttings.
19. An overview of Syrach-Larsen's career is given by Bent Soegaard 1968.
20. Williams 1999, pp. 479–80.
21. Zobel and Sprague, 1993; Burdon and Libby 2006.
22. Zobel and Sprague, p. 53.
23. White and Byram 2004.

24. Eldridge *et al.* 1993. The genus *Eucalyptus* was split in the 1990s into *Eucalyptus* and *Corymbia*. The plantation species of interest are in *Eucalyptus*.
25. The Second World Eucalyptus Congress was held in Brazil in 1961.
26. Eldridge *et al.* 1993, p. 28.
27. White *et al.* 2007. pp. 471–2.
28. Kanowski and Borralho 2004.
29. White and Byram 2004.
30. Wu *et al.* 2007.
31. Wu *et al.* 2007.

9.

FERTILISING

The foresters had to change their attitudes to fertilisers if they were to meet the ever-rising demands for industrial wood. Instead of seeing their forests as sources of nutrients for farms or chemicals for industry, they had to see them as tree farms that could flourish best if they fertilised them. Like breeding better trees, it proved to be a complex and often uncertain business that took them into the new realms of forest science described in this chapter. The scope of their investigations ranged from forest nurseries to natural forests, from heaths to industrial plantations and even to therapeutic attempts to counter the effects of air pollution. To undertake them, they needed the evolving concepts of soil science and plant physiology that were being pushed more by farmers than foresters.[1]

It was the insights of the German chemist, Justus von Liebig in the 1830s and 1840s that marked the start of a modern understanding of plant nutrition. It epitomised the 'metabolic rift' between the almost closed ecology of the village-rural society, whose fields and gardens were fertilised with dung, and the urban-industrial society that imported its foodstuffs and exported its waste as effluent.[2] Liebig showed that plants obtained their nutrients from the mineral elements such as potassium and phosphorous in the soil and from the atmospheric elements of carbon and nitrogen. If the nutrients could not be returned to the soil through dung, they would have to be returned artificially as fertilisers: the balance must be maintained or the soil's fertility would be exhausted. It was a powerful idea, extended but not replaced when it was understood that soil was also created very slowly by natural processes.

European and American farmers turned chemical knowledge to practical effect by importing nitrogen-rich guano and by buying the artificial fertilisers: superphosphate that was being made from the 1840s, and nitrogen fertilisers from the 1920s. Meanwhile the knowledge of plant nutrition advanced to distinguish between six 'macronutrients' that plants needed in considerable quantities, seven 'micronutrients' needed in small amounts and other 'traces' found in plants. Different plants were found to grow best if they had different proportions of the various nutrients. The idea of balance was further extended as scientists realised that the proportions of the nutrients added as fertilisers had to be controlled – too much of one might cause problems with others – and be prescribed according to the type of soil.

Complicating matters further was the realisation that bacteria and fungi in the soil affected growth and were critical for some plants. A major discovery, made in the 1880s was that bacteria in nodules on the roots of legumes could take nitrogen from the air and convert it to a plant nutrient. This ability to 'fix' nitrogen was later found to be associated with some other families of plants. Although farmers used this understanding to good effect in rotational cropping, foresters could not readily use it in practice, even though a few trees, such as acacias and some conifers, were leguminous. However the discovery, also made in the 1880s of how fungal micorrhiza on the roots of plants gave a mutual, or 'symbiotic' assistance to water and mineral nutrition did prove important to foresters. When they established plantations of introduced pines in other lands, they needed to ensure that some micorrhiza were in the soil. For example, in the 1920s foresters in Western Australia found that they had to inoculate new nursery sites with soil from under an established plantation or a fungus slime ball before maritime pine seedlings would grow well.[3]

The depletion of nutrients from the forest was not as easily recognised as that from farms with their annual crops, but it was eventually realised that several causes were linked to agriculture. Swidden or 'slash and burn' farmers relied on the nutrients released by burning the debris on patches they cleared in the forests. They moved on to a new patch when their cropping had exhausted the old; only returning to repeat the process decades later when the forest had re-grown. Extensive grazing areas of former forest or high moorland were burnt annually in Europe, which gradually depleted their nutrients, and similar regimes were applied in other parts of the world. In many of the cold parts of Europe and Asia, nutrients were literally carried from the forests to the farms in the leaves that peasants raked from the forest floor not only to provide bedding for their animals over winter, but also, when mixed with the dung of these animals, to fertilise the fields in spring.[4] When there were not enough leaves, spruce twigs were chopped and added to the byres. Nutrients also left the forest in the wood, although there were less of them per kilogram than in the twigs and leaves.

The obvious changes when hardwood forests were converted into spruce plantations alarmed the foresters. The fact that they were 'monocultures', plantations of a single species, worried them from the end of the nineteenth century; it still does. Instead of the soft, friable 'mull' humus found under the broadleaved hardwood trees, they feared that the harder, acidic 'mor' humus accumulating under the conifers they had introduced meant that the soil was deteriorating.[5] The forests are less healthy as they suffer from attacks of bark beetles and are more liable to damage by storms and snow.[6]

Early trials

Some foresters were keen to see if the way farmers were increasing their yields with fertilisers could be repeated in their forests. The first to try was probably the French forest owner and politician Eugéne Chevandier in 1847. He added artificial calcium, sulphur and nitrogen fertilisers, wood ash and lime to beech, pine and spruce trees. Five years later he reported that growth had been boosted in some by up to forty per cent.[7] Over the next fifty years experiments in France, Germany, Austria, Belgium, The Netherlands, Switzerland, Sweden and other countries showed promising results.[8] Potash, nitrogen fertiliser and lime were tried in many of them and 'basic slag', a waste product of steel-making, proved a cheap and useful fertiliser for its phosphate and liming effects.[9]

The forest scientists interested in fertilisation met to exchange information and design further experiments. For example, the German Association of Forest Research Stations had fertilisation on its agenda from 1904 and IUFRO took it up from 1910. The next year, the Austrian Forest Research Institute in Vienna was able to publish the first findings of its experiments into fertilising young black pine forests. With so much work going on, the German Agricultural Society had a Special Committee to review progress in 1932.[10] The scientists had been able to demonstrate substantial responses to various fertilisers with forest trials in several species, sites and countries. However, they had not analysed the soil for most of their trials and their knowledge of soil chemistry and the physiological effects of fertilisers was limited by the state of soil science at the time.

New plantations

Foresters introducing species to new lands experimented with fertilisers when they found that some of their trees did not grow well. This was particularly the case when they tried to imitate the text-book exemplar of success, the plantations of the Les Landes region of Southwest France (Chapter 6). There, the plantations of maritime pine had been established on dunes and heaths;

in Australia, New Zealand and South Africa foresters tried to establish some of their plantations of radiata and maritime pines on coastal sands, too poor for agriculture, only to find that mysterious disorders – multiple leading shoots, pale colours, dead tops, thin crowns – appeared as the trees grew. The European literature had reported correcting abnormalities in young pine trees with magnesium as early as 1909 so, once they had ruled out fungal disease, they investigated nutrients.[11]

The lead was taken by Bill Stoate in Western Australia from the 1920s. He set new standards. First, he took samples of soil from many small trial plantations that had been established – only some of which showed disorders – and had them chemically analysed. Australian soils commonly have low levels of phosphate and Stoate showed that maritime pine needed at least 150 parts per million in the soil to grow satisfactorily, and that the more demanding radiata pine needed 400. For operations, this showed that failures could be avoided by analysing the soils before deciding where to plant.

The next step was to test for responses to fertilisation. Stoate tried adding not only superphosphate, but also the micronutrient zinc after he saw that some trees growing near to old wire netting fences lacked the disorders: the minute traces of zinc that had leached from the wire had been enough. Stoate designed and analysed his experiments according to the statistical methods laid down Roland Fisher that had only been published in 1925. For example an experiment to test the effects of cultivation, fertilisation and their combination on the growth of young maritime pine used twenty different treatments, repeated three times in a 'randomised block' design.[12]

Stoate designed experiments in different parts of Australia that provided a basis for correcting deficiencies on older plantations and for establishing them through their first two or three years of life, although his insistence on soil surveys prior to planting was not always followed as carefully as it might have been.

Forests

Although they had these many practical trials before them, and accepted fertilisation in some young plantations, few foresters took fertilisation into their general practice. The costs were certain, the financial rewards less so and for many it was an artificial step too far away from the silviculture they had developed for the natural forests. It was only the prick of necessity that altered the situation. After World War Two, the science, or enough of it, had been done to show them that fertilisation offered a possible way to re-

Box 9.1. Theodore Norman 'Bill' Stoate (1895–1979)[13]

'Bill' Stoate

Courtesy of Greg Strelein

'Bill' Stoate, as his fellow foresters called him, was both an Australian forest manager and a research scientist. Like so many of his generation, his life's course was broken by war. He interrupted his forestry course in the University of Adelaide in 1915 to enlist in the Australian Army. In 1917 he was gassed and shell-shocked on the front in Belgium, and was returned to Australia where was discharged on medical grounds. His hair had turned prematurely white. He regained his physical health and completed his course. After working in New South Wales he transferred to the Forests Department of Western Australia in 1922 to assist the only other professionally trained forester then working there, S.L. Kessell. They had to deal with the daily management problems as well as to investigate the scientific ones. As more foresters were trained and joined them, Stoate could spend more of his time on research.

He married in 1924, but was devastated by grief when his wife died in 1930 leaving their two young sons. To help his recovery, he was awarded a scholarship for a year of post-graduate study in the Imperial Forestry Institute in Oxford. On return, he concentrated on plantation nutrition problems. As the country's foremost expert, he was seconded to South Australia to investigate similar problems in 1939–1940. He returned to Western Australia in 1941 to run the Forests Department until 1953, when a new minister decided not to renew his appointment.

Stoate's thirty years of nutrition research was recognised by a doctorate from the University of Adelaide awarded in 1953. It stood him in good stead when he was recruited as by British Columbia's largest company, MacMillan Bloedel as their consultant silviculturist to investigate the use of fertilisers in their forests. It was a relief from petty politics, and an exciting change from correcting problems in young plantations to promoting growth in natural forests. His work there led to a lectureship at the University of British Columbia and an appointment as a professor at Washington State University in Seattle and its nutrition cooperative. His years there provided a fruitful final stage to his career, one in which he happily helped Australian post-graduate students at the university. He returned to end his days in Western Australia.

spond to the intense demand for timber for the reconstruction of European cities. Germany, Austria, The Netherlands, Norway and Finland all started programmes, but it was Sweden, which had stood aloof from the war, that could afford the largest one.[14]

The European lead was taken by the largest non-government forest owner in Europe, Svenska Cellulosa Aktiebolaget (Swedish Cellulose Company or SCA) in northern Sweden.[15] It not only started extensive experiments in the 1950s to see whether forest stands of Scots pine and Norway spruce would respond to nitrogen fertilisers – they did – but also developed the technology to apply them by air. This moved the possibility of fertilising standing forests from experiment to practice. By the mid-1960s its trials had shown that nitrogen fertilisers boosted growth by forty to fifty per cent for up to five years, after which it faded to the pre-fertilisation level. It was enough for the company: if it fertilised, it could increase the amount of wood it cut from its forests within five years, provided it integrated its fertilisation and logging plans. By the late 1960s the company was dropping 60,000 tonnes a year of the nitrogen fertiliser urea on to the areas it targeted.

Meanwhile interest in fertilisation was increasing in North America. In the 1950s laboratory experiments and field trials were established in the Pacific Northwest region of the United States and in Canada, where the MacMillan Bloedel group of companies brought Stoate from Australia to start trials on stands ranging from ten years of age to mature trees. The trials were focused on improving the growth of Douglas fir for timber production, but some research also assisted farmers growing Christmas trees.[16]

The precedent of American companies cooperating in research led by universities, set in the 1950s for breeding, was followed for fertilisation. In the Pacific Northwest a Regional Forest Nutrition Research Project was started in 1969 under the aegis of the University of Washington where Stoate had become a professor.[17] Extensive areas of the magnificent old-growth forest of Douglas fir had been felled over the previous century for the region's large timber industry. In their place was a secondary forest on which the long-term future of the industry would have to depend. The productivity of these younger stands could only be increased by thinning or fertilising them; both had to be tried.

Some of the world's largest timber companies, such as Crown Zellerbach, Weyerhauser, and MacMillan Bloedel, owned extensive forests in the region and had started their own research projects. Although these and other experiments had shown growth responses to nitrogen, they had not been tested across the wide range of soils, sites and ages in the region; nor had the interaction between fertilisation and thinning been evaluated. Such a large experiment, involving 1,900 permanent plots across 270 locations in the US States of Washington and Oregon could only be done by cooperating. The researchers found that stands up to about sixty years of age responded well to fertilisation. Their most important findings were linked to the change in silvicultural practice; instead of relying on the natural regeneration that had produced many of the older second growth stands, the cut-over areas were also being planted. The best responses were obtained if these younger stands were both thinned to a desirable stocking and fertilised one or more times. The research led to about 50,000–55,000 hectares being fertilised a year in the Pacific Northwest by the 1990s. A similar cooperative programme was started in British Columbia in 1971, organised by the Ministry of Forests, and in 1981 a programme for the drier inland ponderosa pine forests was started under the aegis of the University of Idaho.

Industrial plantations

In the Southeast United States fertilisation cooperatives were started in 1967 and 1969, centred on the universities of Florida and North Carolina, which already had well-established breeding cooperatives. There had been a decade of research into the way nutrients 'cycled' between the foliage, roots and soil, and into the response of loblolly and slash pine stands to fertilisation with nitrogen and phosphorous. The responses differed across the region's soils; some of the best were on the 'old-field' sites that had reverted to forest after decades of intensive cropping for cotton and tobacco during which much of their top soil had been lost. Planting these lands was central to the expansion plans of forest products companies. As with breeding, the cooperative structure facilitated research and extensive trials on company lands spread across the region's soil types. Fertilisation was adopted first for new plantations and only later tried on older stands. By the 1990s, about 40,000–45,000 hectares a year were being fertilised annually in the Southeast.

When plantations matured, they were felled and replanted. Their foresters watched anxiously to see if the new crops would grow as well as the first; the spectre of declining yields reported from the Black Forest haunted them. In the 1960s it seemed real in South Australia when a second rotation of radiata pine was found to be producing thirty per cent less than the first.[18] The foresters there shifted the focus of their fertiliser research to the problem and showed that intensive regimes that balanced the proportions of macronutrients to the plants' needs and included several micronutrients could counter the effects. They also showed that it was preferable to avoid burning the debris left after logging, as nitrogen was lost in the process. Instead very heavy rollers were used to crush it and incorporate into the soil. Generally, the problem of possible declines in the productivity of second rotations has been more than offset by the larger gains made by breeding, weed control, cultivation and fertilisation.

Fertilisers are now used in plantations all over the world, mainly in their first years.[19] The responses to fertilisation, and hence the economic benefit, can be quite variable and the foresters had to adjust the type and amount of fertilisers according to the soil and other factors. Chemical analyses of the foliage could help them estimate the requirements at particular sites, but broader assessments were needed to plan operations across large plantations.

Therapy

Foresters in Central Europe became alarmed in the 1970s when they saw fir trees dying and the older needles of spruce trees turning yellow and falling off. The effects seemed worst in the highest forests, the very forests that were most important for protecting the lower lands. Once early observations from Poland, North Germany, the Czech Republic and Slovakia were reported foresters elsewhere examined their forests with worrying results.[20] Air pollution was suspected: the rain was becoming more acid, buildings as iconic as the Parthenon were being damaged by rain and smog; and emissions from British factories were making lakes in Norway too acid for fish to breed. Local effects on trees had long been known, and forest ecosystems are resilient to a certain extent before the decline becomes visible, but this was on an entirely different scale; some sort of general 'forest decline' was feared. It demanded national surveys and intense investigations.

The forest scientists gradually realised that the decline in the forests' health was due to several stressful factors combining. Chemists showed that the pollutants were complex mixtures with carbon dioxide, methane, ozone, nitrogen, sulphur and volatile organic compounds, and photo-oxidants, produced high in the atmosphere by the action of sunlight, all involved. It was a complex problem to unravel, but the forest scientists managed to find out that the plants' sensitivity varied by time and season and that long-term exposure, even in low concentrations, had toxic effects that could weaken and damage forests, particularly sensitive ecosystems such as mountain forests.[21]

The foresters also needed a therapy to restore the forests' health. It was difficult to think of one for such widespread and obscure conditions. However, fertilisation offered some prospects: needle loss was known to be a symptom of magnesium deficiency, acid soils could be limed and resistance to most sorts of damage could be improved by fertilising. Although some researchers conducted nutrition experiments and investigated the soil, they could not devise treatments that could be applied economically on an operational scale. With all their knowledge from one hundred years of forest science, they could not prevent the forests being increasingly endangered.

Meanwhile extensive surveys were made, co-ordinated to some extent by international organisations.[22] As they and other investigations progressed, the complexity of the situation became increasingly apparent. The symptoms were found not only to be variable, but scientifically uncertain because change could not be proved decisively without previous 'base-line' measurements. There were many factors at work in different places and at different times: drought, past silviculture and logging, pathogens. Although there were significant problems in some places with some species, there was little evidence for widespread decline.[23] Moreover, by the 1990s, international concern about the health of the forests changed from the effects of air pollution to the broader concerns about long-term climate change described in Chapter 14. Nevertheless, they should not be forgotten, because thousands of hectares of forests in the former East Germany, Poland, the former Czechoslovakia and other countries were very seriously damaged and the sensitive mountain forests in Europe continue to be weakened. World-wide, the long term exposure to oxides of nitrogen and ozone, alone or combined with other pollutants such as heavy metals or dust continues to affect forest vegetation.

The foresters adopted fertilisation and genetic improvement, described in the previous chapter, as standard practice in much of the Pacific Northwest, the Southern United States and large industrial plantations elsewhere. These practices were accompanied by advances in cultivation, the chemical control of weeds and automated logging. Although none of these are covered in this book, they contributed to overall process of increasing and intensifying the production of wood to suit large-scale industrialisation. Equally important were the advances made in planning, as described in the next chapter.

Chapter 9

Notes

1. Overviews of the history of forest fertilisation are given by Tamm 1968, Baule and Fricker 1970 and Binkley 1986.
2. The term 'metabolic rift' was developed by John Bellamy Foster (2000) from Marx's insights. Liebig's *Organic Chemistry in Agriculture and Physiology* was published in English in 1840.
3. Carron 1985, p. 163. The micorrhizal fungus was *Rhizopogon luteolus*.
4. Tamm 1968.
5. Wiedemann, 1925a, 1925b.
6. Brandl 1992.
7. Raudiére 1930.
8. Ebermayer 1891.
9. Schwappach 1911.
10. Evers 1991, Baule and Fricker 1970, p. 7.
11. Baule and Fricker 1970, p. 6; A fungal disease of *Diplodia pinea* was the original suspect.
12. Stoate and Lane Poole 1938.
13. Information drawn from Mills 2002.
14. Tamm 1968.
15. Hagner 1966.
16. Gessel 1968.
17. Chappell *et al.* 1991.
18. Keeves 1956.
19. Some examples include: phosphorous and potassium for Sitka spruce in Wales; nitrogen and lime for Norway spruce in Denmark; phosphorous for eucalypts in China; nitrogen for the same in Australia; and phosphorous and potassium for eucalypts in Portugal, or for mahogany in Indonesia, for example.
20. Evers 1991, Innes 1993 and the papers in Schlaepfer 1993.
21. Academy of Sciences [Austrian] 1989.
22. Major bodies were: the International Cooperative Programme on Assessment and Monitoring of Air Pollution Effects on Forests of the UN Economic Commission for Europe; the Commission of the European Communities; and the IUFRO Task Force 'Forest Decline and Air Pollution', later enlarged as 'Forest, Climate Change and Air Pollution'.
23. Innes 1993.

10.

PLANNING

The foresters developing new resources for industrial expansion faced problems different from those on which forest science had been founded. Rather than having to regulate the existing forests so that their yield could be sustained in the long-term, they had to build up the supply of wood for new mills or exports in a hurry. One way was to reach further into the tropical and mountain forests with more and more powerful and ingenious machines in an engineering history that is outside the scope of this book. The other way was to develop industrial plantations. They could breed better trees and grow them faster with fertilisers, but it was an expensive business and they had to manage them efficiently. How they planned to do so is the subject of this chapter.

The drive for industrialisation as the path to development enlarged the economic scale on which foresters had to make their plans. Rather than planning each forest from the perspective of an individual forest owner selling wood in a large market, the foresters had to plan their plantations as part of particular developments. Their existing theories did not consider this, although some foresters had experienced it in practice. This change from individual forests to regional or corporate projects meant that they had to consider effects beyond the forest boundaries. This was far from easy because they were not the sole decision-makers and what they planned for their forests and plantations had to fit larger plans.

The largest plan of all was the UN's Decade of Development (1960s). It was intended to stimulate the economies of the developing countries with investment aid from the advanced, industrialised ones. Part of its economic rationale lay in the 'input/output' model of a national economy.[1] It measured how the effects of an investment in one sector might 'multiply' as they spread backwards and forwards through economic linkages to others. Jack Westoby in FAO's forestry division used the model to show how stimulating the forest industries, particularly the pulp and paper industry, could be effective in promoting growth in low-income countries.[2] It provided foresters with an argument for investment when a strict economic analysis of a stand or a plantation on its own might show it as unattractive. On one hand this risked boosting uneconomic projects that would eventually fail, while on the other hand it demonstrated that a single criterion – maximising individual profit – was inadequate for planning something as complex as a forest region. The forest planners needed new methods.

Their new methods came from a bundle of mathematical and statistical techniques that military planners had developed for the complex problems of making decisions during World War Two. Afterwards they were applied to business and industry and became known as 'operations research'.[3] It was a new approach to providing a quantitative base for making decisions that has since blossomed under labels of 'management science', 'decision theory', or 'systems theory'. Two features of this diverse field are the construction of mathematical models to analyse the systems being studied and the evaluation of numerous options before making decisions. The latter was particularly important to the way foresters thought about making their plans; rather than work from their foundation principles of optimum rotation and sustained yield, they should generate and choose from a wide range of options.

The digital transition

Operations research was an early part of the 'digital revolution' in communication, science, politics, business and almost every aspect of contemporary society. The revolution was as much about the software, the programmes that people used, as it was about the increasing power and speed of the computers themselves.

The digital transition from military and research laboratories to general use was as liberating and exciting in forest science as it was in other fields. Not only could the forest assessors drastically reduce the laborious, boring work of calculating the results of their measurements in the forests – this could take half their time – the forest researchers could use more sophisticated models for analysing their data statistically. For example: they could calculate equations to describe the shape or 'taper' of trees in ways which enabled them to estimate the volume of logs of different sizes quickly; they could finally replace all the tables for calculating the volume of trees and stands with equations; and they could evaluate the interactions between multiple factors in their experiments.[4]

A few forestry agencies were already using computers in research by 1960 but the transition into general planning practice was necessarily slower.[5] The records, maps, and graphs with which foresters everywhere had built up their knowledge in places for over a century were voluminous, but they were on paper. The task of 'cleaning up' the old measurements written by different people over many years to the unforgiving digital standards was so enormous that many records were abandoned. However, all the new measurements were put in digital form.

The forest researchers had to learn new skills. Although they could use standard statistical programmes to calculate their equations or analyse their experiments, they had to write their own programmes to process the forest inventory data and then to assemble it with the equations into models of the stands and forests.

The transition occurred first in the American centres where computer facilities were most available and skills were developed, but it spread elsewhere. By the mid-1970s over half the published studies still originated from the United States and Canada, but others were starting to come from Sweden, the United Kingdom, other European countries, Australia and New Zealand. Much of the early work concentrated on turning earlier graphs of tree shapes, volumes and site qualities into equations. Although most of it was concerned with even-aged conifer stands, a few researchers started to quantify the distances between trees in uneven-aged, mixed-species stands. With such work in hand, researchers could turn their attention to modelling whole stands.

Simulating stands

Foresters had sought to quantify the growth and yield of stands from the foundation of modern forestry. This was essential for making long-term forest plans. In separate analyses, their experiments had sought to quantify how their tending could change the stands. In the digital era they had the possibility of integrating both areas; they could incorporate their silvicultural tending options – spacing, thinning, fertilising, breeding and so forth – in the *same* analysis as the various ages at which they could fell the stands. And they could estimate all the different sizes of trees and logs that could be grown. And if they wanted to, they could add in the discounted cash flow calculations while they were at it. Moreover, they could do so not just for the half-a-dozen standard classes depicted in the yield tables but for the whole variety of conditions across the forest.

Forest scientists built the first digital 'growth and yield models' for even-aged conifer plantations in the 1960s. Some of the leading work was done in the United States by Jerry Clutter who set up a biometrics/operations research group in the University of Georgia with industry funding in

1963. It focused on loblolly and slash pines in the southern United States, with Clutter and others also doing work on South African, New Zealand and Australian pine plantations.[6] The group made theoretical advances that enabled a model's several equations to be statistically compatible with each other so that it could be used in forecasting and they showed how more subtle depictions of biological growth could be made using 'non-linear' equations.[7] An important advance was the ability to show statistically how the frequency of tree sizes in a stand changed with time. Many of the forests in the region were being converted from the old-field sites of naturally regenerated pines for which the group developed models to deal with their variable densities and allow for the natural mortality of the smaller trees that occurs over the life of un-thinned stands.

Following the same cooperative arrangements with large forest products companies that had been used for breeding and fertilising research, the Georgia group was supported by seventeen companies who made their data available and tried the new methods. It became the Plantation Management Research Cooperative and was followed in 1979 by another cooperative, formed under the aegis of the Virginia Polytechnic Institute and State University. The large scale manufacturers, such as those in the southern United States and Sweden, wanted information about the size of the logs that would be coming from their plantations and forests so that they could develop mechanised logging machines and automate the wood handling at their mills. This gave a further impetus to the researchers developing models of tree shapes and the frequencies of tree sizes in stands.

In operations research terms, the growth and yield models were said to 'simulate' the behaviour of real stands, so that the likely effects of possible alternatives could be estimated. For example, some in Australia, New Zealand and South Africa were designed to show the effects of thinning regimes in pine plantations, each varied by the frequency, severity and type of tree removed. The long tradition of growth and yield studies in European research centres continued. It was generally more focused on natural forests with their longer rotations, than on plantations, and was later in making the digital transition.[8]

The growth and yield models of stands provided forest planners with a rational means of estimating the consequences of the various options they had for managing them. Like the old yield tables, they were most reliable where the data used to construct them was strongest, but they could be used – though with less confidence – to forecast the yields likely to be attained at ages or in situations beyond the reach of strong data. Although the planners could generate the likely consequences of many different options for every type of stand, they needed larger models to plan for whole forests or plantations.

Simulating plantations

Many of the world's new industrial plantation schemes were started by governments that hoped to attract industrial developers with the wood resources that they would eventually build up. Even when the plantations were started by companies to supply their own mills, the actual date and size of the mills eventually constructed might differ from their intentions many years before. This created a difficult planning problem.

The foresters could follow their normal forest ideal: if they planted an equal area each year for as long as the rotation – say twenty years for a pulp mill or forty for a sawmill – then they could supply them for ever. However, their concept was severely flawed in three major ways. First, they needed to plant smaller areas until they had proved the trees would grow well and they had built up enough experience to forecast the future yields from a large scheme. In practice, governments and corporations usually funded projects in fits and starts, so that the rate of planting and hence the eventual yield would fluctuate from year to year. Second and most important was that the timing of investments was determined by general economic circumstances and by how particular companies valued the resource and judged their market prospects. The foresters might establish large plantations only to find that there were no mills to take the wood when it was ready. This was particularly difficult with species that needed thinning as the un-thinned stands would not produce the best trees and might be unhealthy. Third, to make new pulp and paper mills profitable, they had to be large; they were 'lumpy' investments and their minimum economic size kept increasing.[9]

This all placed the foresters in two uncomfortable situations: while they were waiting for the mills to be built or to expand they could not sell their wood, but immediately they started they had to produce enough wood run them at full capacity. They had to plan for yields that could flip-flop from gluts to shortages. By circumstance or choice, they had to be able to 'move' wood in their plans from one period to another. If they were constrained by industrial capacity and had too much wood available, they could do little else but delay thinning and felling. They would have still more wood later, although there might be risks of poorer quality or storm damage in some types of stands. In the opposite situation, if industrial capacity was constrained and pressing, they could thin more heavily, or in a different way, or fell at younger ages, but would have less wood later. This was a more risky situation, as they had to ensure that they would have enough for the future and had no reserves against fires or storms. In some places they could fertilise stands, so that they would yield more in a few years without such losses, as in a Swedish practice.[10] Two idealised cases are depicted in Figure 10.1.

The planning problem of how to 'move' wood from one period to another could not be solved from principles alone and there was no simple way of doing so. The stand simulation models could estimate the future yield of each stand for the thinning and felling options, but putting them all together to show the total yield from a whole forest or plantation at different periods in the future was a formidable computational task if done manually.

Plantation and forest simulation models started to be developed to do this from the late 1960s.[11] The first models simply repeated the clerical process – of applying a standard yield table to the areas of each age – on a computer but they were improved once the assessments of the stands' quality and density had been 'digitised' and when variable-density growth and yield models had been developed. These were used to forecast the future yields for each stand, for the different types of thinning and felling ages that were possible, typically for the next twenty to thirty years.

Simple 'decision rules' were used to choose which stand and which option to use until the mill demands were met for every year. They might be as simple as: 'clear fell the oldest stands first'; or they might be 'use heavy thinning for the first period, moderate thinning afterwards and clear fell the oldest stands first'. Although these models were conceptually simple, they could be used to generate several plans by changing the decision rules and progressively refined until a reasonable way of meeting demands had been found. Each iteration needed to be 'run' separately on a computer which limited the number that the early planners could make; between five and twenty runs spread days apart would have been typical for many. Although it was drawn out, the process of making iterative simulations of a large plantation scheme proved effective in practice and its conceptual simplicity made it relatively easy to add features such as discounted cash flow analyses, or to revise the plans after fire or storm damage.

Although these iterative simulation models could be used to find an acceptable plan for an enterprise, they did not necessarily find the best one; for that a different form of model had to be developed.

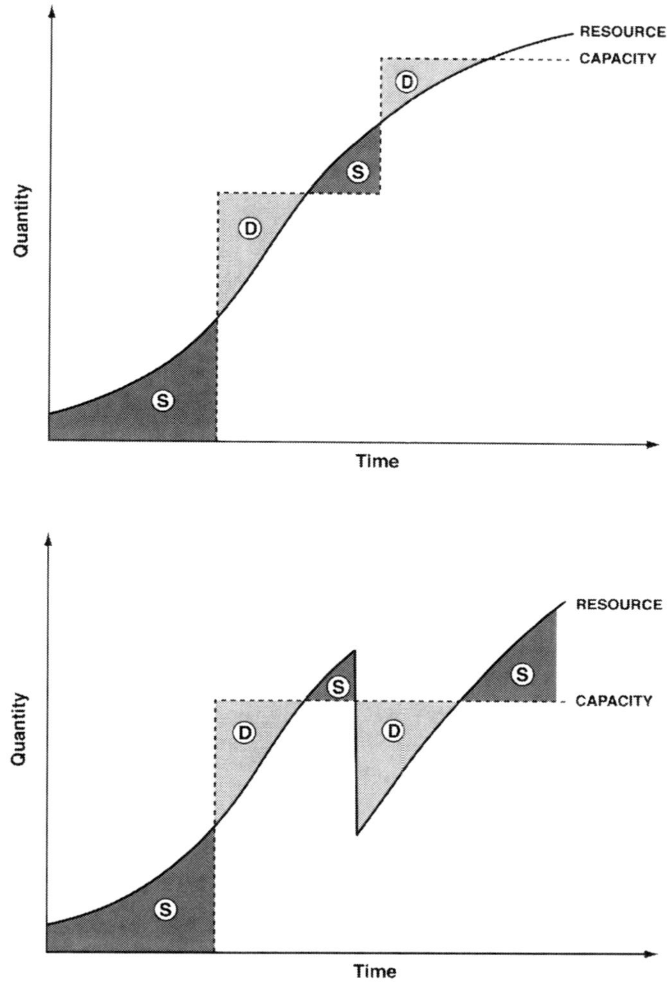

Figure 10.1. Idealised diagram of the plantation-industrial system showing periods when:

S = Surplus wood available from standard silviculture is surplus to demand; the plantation/industrial system is constrained by capacity.

D = Demand demand is greater than the wood available from standard silviculture; the plantation/industrial system is constrained by supply unless rotation and/or is changed.

Top case: steady expansion of resources available and stepwise expansion of industrial capacity.

Bottom case: supply is interrupted by catastrophic fire or hiatus in planting.

Optimising plantations

In 1968 Clutter's group in the University of Georgia produced the first model that used linear programming to analyse the forest-planning problem for industrial plantations in the Southeast United States. In forest planning it was a revolutionary change in several ways: rather than the planner making repeated iterations of a simulation model, the Georgia model made all the iterations within a computer; rather than consider only a few combinations of silvicultural and felling options, it considered many thousands; rather than the planner selecting one of the iterations that was judged to be satisfactory, it selected the single *best* one mathematically — it optimised.[12] It also provided economic insights into the relative merits of the selections it calculated.[13]

The development of the Georgia model crystallised a reappraisal of the founding principles of forest regulation.[14] The most incisive was the rejection of achieving the normal forest ideal of sustained yield as the objective of planning, in favour of the economic objective of maximising an owner's benefit. Although this simply extended the well-established discounted cash-flow methods for single stands to whole enterprises, it envisaged management as a process of continual transition, rather than as a struggle for an ideal – today it is called 'adaptive management'.[15] Viewed in one way, it made little change to the way foresters had always bent their plans to what they could achieve and had revised them periodically; viewed another way, it was an assertion of economic values. The reappraisal also recognised that the traditional methods of site qualities and standard yield tables could not use all the rich detail becoming available digitally from the inventories and growth plots. With greater recognition of complexity, foresters could have much greater flexibility in deciding when to thin or fell their stands than the traditional methods could allow.

The Georgia model was taken up quickly by its cooperating companies who were already using it by 1969 to plan four million hectares of plantations. However, the pace of development elsewhere was limited by the availability of sufficiently powerful computers, specialist forest planners and qualified programmers. By the end of the 1970s, similar linear programming models had been developed in Australia and other countries where they proved remarkably adaptable to the needs of particular enterprises.

The difference between the simulation models and the optimising models proved to be more one of detail and insight than in the way they were used. The objective of maximising profit in the optimising models oversimplified the decisions that enterprises made. Companies needed to consider

Box 10.1 How the optimising model works

The ideas of the model were quite simple and can be explained in the case of making a plan for a hypothetical plantation of a single species grown to produce only pulpwood for a single mill. The plan was to be made for the next 25 years.

Each of the plantation's 300 stands had data defining its area, age, site quality and stocking. The stands were not to be thinned and could be felled at any age between 11 and 30 years so that there were 20 options for how each (fell at 11, or at 12, 13, ... or at 30 years of age). Every time a stand was felled it had to be re-planted.

The stand model was used to calculate the yield expected from each particular stand for each of the options, which meant that 6,000 calculations were made. As the costs of planting and maintenance and the value of the wood to the mill were known, the net present value of each option for each stand was calculated at the same time. All the yields were then placed in a table with a line for each of the 25 calendar years in which they would occur and the values were placed in their own line.

Whichever option was finally chosen for each stand, their total in each year had to equal the amount needed at the mill. This meant that there were 25 equations, one for each year, that had to be solved together. There might be other 'constraints' that needed additional equations, such as that the amount that could be spent each year had to be less than the budget available. The total of the net present values of the options finally chosen made another equation that calculated the total value to the owner - the 'objective' of the plan.

The foresters would have learnt at school how to solve 2 or 3 simultaneous equations with 3 or 4 variables in each, but 25 or 50 equations with 6,000 variables was an entirely different matter that only the linear programming method in a computer could resolve. The great strength of the method was that it could test every possible combination of options and choose the one that gave the maximum total benefit.

strategies against competitors, cash flow and other opportunities; while governments needed to consider social factors and political consequences. Moreover, setting the true value of the wood to use in these models was fraught with difficulty. The prices that companies used for internal transfers between their plantations and their mills were artificial and the prices at which governments sold wood were often set politically. In practice, the more powerful optimising models were used to create several plans, each one being a 'best' solution to different organisational factors.

The forest planners improved the models considerably once they could assemble their maps and assessment information in geographic information systems. These systems used statistical methods of spatial analysis that in the forest sector had been developed for the problems of conducting continuous forest inventories for large regions or nations. Bertil Matérn in Sweden had taken a leading role in this. Models that could include a spatial analysis became increasingly more important economically as the level of investment and the intensity of management increased.

An example from Brazil can illustrate the way in which the advances in tree breeding, fertilisation and modelling have been brought together in operational planning in the twenty-first century. Brazil embarked on a spectacular programme of plantation expansion in the 1960s with an incentive scheme of tax concessions that lasted for twenty years and was designed to attract foreign investment in the pulp and paper industry and provide charcoal for the steel industry. Over two million hectares of eucalypts and pines were planted for industrial use, enabling Brazil to become the fourth largest producer of paper pulp in the world. Their success was due to the favourable climate and soils, combined with research on species introductions, breeding and fertilisation. The Brazilian foresters could grow industrial wood faster and on shorter rotations than their competitors on the world market for paper pulp. For example, they could claim to grow eucalypts on seven-year rotations to produce 44 cubic metres of wood a year, whereas the Swedish foresters grew spruce on seventy to eighty year rotations to produce four cubic meters a year.[16] They needed intensive planning methods to get the best out of their scientific advances and natural assets.

Our example comes from International Paper Company's plantations in the State of São Paulo where they supplied the pulp mill built in 1996. International Paper, now the world's largest pulp and paper company, had been an early member of the tree breeding, fertilisation and planning cooperatives in the southern United States. It brought an intensive management and research culture to developing its Brazilian plantations.[17] Timor mountain gum, a eucalypt, grew particularly well; provenances were selected, tested and the best were reproduced as clonal plantations. Fertilisation with nitrogen, phosphorous, potassium and calcium increased growth dramatically, but the response varied across the three regions and soil types of the plantations that could produce between 56 and 64 cubic metres of wood a year. As the cost of fertiliser increased, but the world price of pulp did not, the planning problem of where to apply the fertiliser to best effect became more impor-

Box 10.2. Bertil Matérn (1917-2007)

The Swedish forest statistician, Bertil Matérn was born in Gothenburg in 1917. He studied at Stockholm University and worked there from 1941 in the University's Institute of Insurance Mathematics and Mathematical Statistics. In 1945 he was hand-picked by Professor Manfred Näslund to join the National Forest Research Institute and work on the country's forest inventory. This was an important appointment because the statistical design of a national inventory, with its plots spread across the entire country, is critical to its cost and the validity of its results. Matérn spent a year of further study in the United States at the Universities of Iowa and North Carolina and in 1947 married Carin Berglund, who had also studied at Stockholm University.

Bertil Matérn

Downloaded from *Skogen*, 6 February 2013.

Matérn was deeply involved in the national forest inventory and from 1951 worked as a specialist in the Royal College of Forestry, part of the University, that trained foresters. He was awarded his doctorate in 1960. In 1962 the National Forest Research Institute and the forestry school merged and an Institute of Forest Mathematical Statistics was established within it, with Matérn appointed as professor in 1963. In the 1970s, the University relocated to three campuses at Umeå, Uppsala and Garpenberg, which involved him in frequent travelling until he retired in 1981.

Matérn was a pioneer in the development of spatial statistics and their application to forest problems. His 1960 doctoral thesis is a milestone in the area and was reprinted in 1986.[18] One chapter is known as the 'Matérn process'. It is frequently quoted and increasingly applied.

Matérn had leading positions in several international bodies such as IUFRO, and in forest associations. He spoke several languages and was hired both at home and abroad for investigation and expert missions in Austria, Mexico, and East Germany.

He retired in 1981 and died in 2007.

tant. A new way of measuring the responses was devised by installing 130 pairs of plots (one treated, one untreated) spread across the whole 34,000 hectare plantation. The results, when analysed statistically, showed that soil type, age and region were most important and that there was no difference between the two major clones that were planted. These results were related to the geographic information system to make fertilisation response maps for the forest managers.

Convergence and divergence

In the processes of increasing and intensifying production to boost industrial development, described in this and the previous two chapters, the traditional distinction between natural forests and artificial plantations was progressively diminished: both were becoming tree farms for industry. Natural regeneration was replaced by the more reliable planting; the trees were genetically improved; their spacing was controlled; they were fertilised when they were planted, and sometimes when they were older; and their weeds were controlled with herbicides. There were fewer people in the forests, as labour-intensive work was replaced by capital-intensive, mechanised work; and the advanced planning systems helped arrange such changes in the interests of large corporations. This convergence between forests and plantations can be seen as another expression of the metabolic rift between country and town, between nature and industry.

Convergence was neither total nor unchallenged. It was expressed most strongly in the conifer forests of the Southern United States, the Pacific Northwest and Scandinavia. It was weaker in the spruce areas of Central Europe and weakest in the hardwood forests with their mixed species and ages. It was denied in that strand of silviculture that sought to work in harmony with nature.

Stand models and advanced planning systems were also developed for the hardwood forests but at a slower pace than those described in this chapter for the industrial plantations. The forests, with their mixtures of species and ages, were more complex, it was harder to translate their records into digital form, they generally grew more slowly and there was less economic pressure to develop them. However, it was to these forests with their multiple uses and values that forest scientists turned their attention.

In the 1970s, the vigorous pursuit of development that had characterised forestry since the 1950s was challenged by the equally vigorous pursuits of environmental and social values. In the process, the hopes and visions of a desirable forest future became split and the paths of forest science diverged further, as described in the next section of this book.

Notes

1. It was quantified in a matrix in the 1950s by the Russian-American economist, Wassily Leontif, who was awarded a Nobel Prize in 1973.
2. Westoby 1962.
3. It became a field with its own journals. The *Journal of the Operations Research Society* was published in Britain from 1950, *Operations Research* was published by the Operations Research Society of America from 1952 and similar journals were published subsequently in other countries.
4. An inspection of the early issues of *Forest Science* shows that the digital transition affected every area of forest research but was taken up most strongly at first in these sorts of topics.
5. Jeffers 1961 reports that, by 1960, computers were being used by forest organisations in Australia, Canada, Sweden, Switzerland, the United Kingdom and the United States, and to a lesser extent in Japan, the Netherlands and New Zealand. They were also used in Austria and Germany during the 1960s and are now universal.
6. One of the authors (J.D.) from Australia joined the group as a Research Associate for a year in 1969–70.
7. The advances made by the group and others are covered in Clutter *et al.* 1983.
8. Pretzsch 2009 gives a comprehensive treatment of growth and yield studies with a focus on European research.
9. The minimum economic size for a pulp mill varies according to the type of process and whether it is integrated with paper production on site. An indication of the trend can seen by comparing the average size of pulp mills (all processes) which was about 80,000 tonnes per year in the 1960s, with the size of new pulp mills that are now built to produce 500,000 to over one million tonnes a year.
10. Described in Chapter 9.
11. Some are detailed in Dargavel 1976. Some models were also built in Australia to simulate the plantation-industrial system across the country and for individual states.
12. Clutter *et al.* 1983. The large planning models were named, often with acronyms. The Georgia model was MAX-MILLION.
13. These marginal values could be useful to the planner in helping to understand the economic functioning of the plantation-industrial system. For example, if a particular option had a very low marginal value, then it could be used with negligible loss if there was some operations reason to do so; if it had a high value, then it was clearly very undesirable.
14. Ware and Clutter 1971. The founding principles are discussed in Chapter 4.
15. The discounted cash-flow methods are discussed in Chapter 3.
16. Cited by BRACELPA 2011. The Brazilian competitive position in growth rates is being challenged by countries in Africa and Asia.
17. Ferriera and Stape 2009.
18. Matérn 1960.

DIVERGENCE

The whole structure of forestry, its science, its institutions, its foresters, its scientists, its mentality, its vision of sustaining the forest's yields for ever, was challenged by a rising concern for the environment, not just of forests but everywhere. Forestry had to change.

There had never been a time when humans, or at least some of them, had not been aware of their effects on the natural world, but in the late twentieth century their attention reached a new level. Encouraged by an environmental movement that blossomed from slender beginnings, 'the environment' became an entity in the popular imagination in a way that it never had before. 'Environmentalism', previously a scientific theory in the debates about evolution, became the social and political drive to protect the natural environment from human harm.

Addressing the problems scientifically needed insights from several disciplines. In the 1960s and 1970s environmental science emerged as a broad field that summoned whatever disciplines and specialities it needed for a problem: ecology, chemistry, geology, forestry, atmospheric science or rural sociology might all be called on, for example. Even for forest-related problems, forest science was only one of many disciplines that might be needed. Environmental science saw each problem as a system of trends and flows of materials or energy whose effects and interactions could be traced, but it had to overcome the separate ways each discipline constructed its knowledge.

The rise of environmental concerns was marked in 1969 by national legislation in the United States that not only required comprehensive environmental impact statements to be made whenever a significant development was proposed, but also allowed for public consultation before it was approved. Its principles were soon followed in other countries and the public demands for information and participation in all development decisions became part of governance, at least in principle. No longer were the foresters in sole charge of making their long-term plans; to their discomfort, they had to listen to other views, many from an increasing number of non-government organisations. The most important scientifically was the International Union for the Conservation of Nature (IUCN), formed in 1948 by a congress of states, government agencies and national and international non-government organisations.[1] The rise of environmental concerns reached the United Nations level with its Conference on the Human Environment in Stockholm in 1972.

The rise of environmental concern was only one of many changes: decolonisation, demands for public consultation and equal rights for women, broadening the concept of development to consider all layers of society and so forth. The United Nations reflected some of these by shifting the emphasis of its Second Decade of Development (1970s) from industrialisation to alleviating poverty. No longer could foresters manage the forests just for the construction and industrial markets; they would have to consider wider social issues, and re-think their foundations of regulation and planning. Then the 1973 and 1979 oil-price shocks coincided with droughts in Sub-Saharan Africa to highlight the importance of fuel wood for both energy and subsistence in poor countries. Much of that came from woodlands and species outside the existing body of forest science. The foresters had to find new approaches for new circumstances.

In the previous section of three chapters, we described the energetic development of industrial plantations and discussed how this process had diverged from the management of natural forests. The environmental and social challenges of the 1970s led to a further divergence in forestry practice and the forest science to advance it. The following three chapters describe the way forestry diverged further into one adapted to environmental concerns, one as a form of conservation free from human use and one for small-scale local needs.

Note

1. It was formed as the International Union for the Protection of Nature and changed its name in 1956 to the International Union for the Conservation of Nature and Natural Resources with the acronym IUCN. In 1992, 'The World Conservation Union' was added to its logo.

11.

BALANCING

Wood, how much there was, its shapes and sizes, how fast it grew, how to ensure the forests did not run out of it and how to make them more productive; these were the concerns that had driven the foundations of forest science, its extension around the world and the development of industrial plantations. But they had never been the only concerns. The forests had too many uses, too many users, too many values for that. Some could be ignored, and the overall balance might be weighted by wood, but most had to be accommodated one way or another, forest by forest.

If we step back to before the modern forest science of this book, we can find a mass of rights for peasants, commoners and townsfolk to use the different products of the forest. Many of these old rights were extinguished, but the needs persisted for what the textbooks called 'minor products' which the foresters had to supply or control. Some, like keeping bees, might positively help the forest; others, like gathering flowers, could occur independently; hunting usually reduced full wood production to some extent; while intensive grazing might damage or even eventually destroy a forest. During the nineteenth century, the way in which the forests affected soil and water, their values for scenery or recreation and how they provided all the various uses came to be called the 'functions' of the forest or, in today's managerial jargon, 'ecosystem services'. Few if any of them appeared new; they only needed to be given greater weight to keep them within forestry's scientific domain, or so the foresters hoped. To do so they had to extend their scientific base of measurement, experiment, modelling and planning to issues beyond wood. How they did so is the subject of this chapter.

Well before the environmental era started in the 1960s and 1970s, foresters had taken two main ways to manage the interactions between these various uses and values of the forests. The nature-based management advocated by several European foresters rested on the philosophical ideal of a society being in harmony with nature. The 1880s and 1890s was also a time when the beauty of forest stands and single trees was being increasingly recognised.[1] The foresters developed the 'control method' of measurement, silviculture and regulation in France, Switzerland and Germany to manage forests with intimate mixtures of species and ages.[2] The core of the theory was that when wood was produced with such methods, all the other functions could be provided without additional and external costs.[3] Meanwhile the American foresters followed a utilitarian precept for managing the US federal forests, declared in 1905 as seeking 'the greatest good for the greatest number', with the forestry addition of it having to be 'in the long-run'.[4] They applied it, case by case, as they gradually introduced systematic management to their varied public forests. Although by the 1930s and 1940s they could cite successful examples of what they had done, they lacked a rational, quantitative way of balancing the different interests to achieve the greatest public good.

This American understanding of 'multiple-use' was taken as the central theme for the Fifth World Congress held in Seattle in 1960. The foresters' objective, it was asserted, was to achieve a high and sustained level of production for the five major uses: wood, water, forage, recreation, and wildlife. As the environmental era progressed, governments pressed their forest services to recognise wider environmental dimensions. The United States set the comprehensive tone in 1976 when it declared that its plans had to 'provide for diversity of plant and animal communities' and ensure that any clear-cutting was 'consistent with the protection of soil, watershed, fish, wildlife, recreation, and aesthetic resources, and the regeneration of the timber resource'.[5] Other governments followed suit: in Australia's State of Victoria, for example, plans had to be 'economically viable in providing wood, environmentally sensitive, sustainable for *all* values for future generations, and assisted by public participation in planning'.[6]

In 1992 such national directions rose to the international level at the United Nation's 'Earth Summit' conference held in Rio de Janeiro.[7] Its world vision of 'sustainable development' was expressed as 'ecologically sustainable development' and a statement of principles for the forests. These were given a systematic structure for the different regions by defining indicators that could be used to measure the progress each country was making under environmental, economic and social criteria.

Grand declarations were one thing; being able to meet them was quite another. The foresters tackled them in stages. At first there were particular issues to be dealt with one-by-one, but by the 1990s their vision of sustaining the forest's yield for ever – sustained yield – had been transformed into one of 'sustainable forest management'. This was seen as finding a balance between meeting increasing demands, as they had always tried to do, and maintaining the biodiversity and health of the forests. It was just as much a vision for the long-term, but it replaced the idea of there being an ultimate goal – the sustained yield from a normal forest – with the idea of adapting the balance when circumstances change. The foresters started from what they knew by extending their foundations of measuring, tending, deciding and regulating. They approached the task by modifying their everyday operations; scientifically through economics, planning and silviculture; and by building on their existing strand of nature-based forestry.

Operations approach

They still had to produce the wood. For the foresters, it came down to the practical question of how to conduct their logging and other operations while also satisfying the environmental issues as they emerged. Soil and water, wildlife and plants, recreation and scenery, sacred groves and historic sites were hardly new concerns in many countries, but they had to be given a scientific base and brought systematically into the way the forests were managed and regulated. Much of what had to be done was specific to particular places and types of forest. Some of it was banal, like stopping logs being dragged through streams or waste oil being dumped in the forest; some of it rested on formal scientific research, like measuring the effects on fish of shading streams or the run-off of silt from logging tracks and roads; but much of it was developed less formally. The issues being considered became increasingly comprehensive and progressively formalised in 'codes of forest practice'.

The first jurisdiction to require a code of forest practices was Oregon in 1971, several other US states followed, as did most Australian states during the 1980s. Ireland was the first in Europe to do so in 2000 and, in order to reach developing countries, FAO prepared a model code for forest harvesting in 1996 that any country could use. Codes varied considerably in the detail they required, whether they applied to private as well as public forests, whether they were compulsory or not and the extent to which they were policed.[8]

The codes of forest practice became more important after 1994 when a consumer-based form of regulation was started. Consumers were encouraged to buy their timber only if it was certified as coming from forests that were managed sustainably.[9] Codes of practice and other criteria were needed to determine whether a particular forest met the standards set by the various certification schemes.

Box 11.1. Forest Practices Codes, some examples

Rivers, Streams and Lakes

Ecologists call the area around rivers, streams, and lakes, the 'riparian zone'. It is the richest part of the landscape for plants and animals. Its vegetation traps the sediment that comes down the slopes, prevents the banks from eroding and stops the streams from silting up. No logging is allowed in buffer strips either side of the water and special care has to be taken to protect a zone behind that.

Jurisdiction	Type	Protective strips
Tasmania	Streams from catchments over 50 hectares, used for town water	10 metre buffer in 50 metre special protection zone
"	Small streams from catchments over 100 hectares	10 metre buffer in 30 metre special protection zone
"	Major streams	10 metre buffer in 40 metre special protection zone
British Columbia	Major rivers over 100 metres wide with fish or used for domestic water	100 metre special protection zone
"	Rivers between 20 and 100 metres wide with fish or used for domestic water	30 metre buffer in 50 metre special protection zone
"	Stream less than 3 metres wide	20 metre special protection zone

Clearcut areas

Clear cutting, or 'clear felling' all or the majority of the trees at one time is the foresters' most drastic treatment and is the cheapest way of logging. They use it for industrial plantations and commonly in the natural forests to regenerate light-demanding species or to create even-aged stands.[10] Removing the canopy, especially if followed by burning, exposes the soil to erosion, destroys the habitat for a time and, to most people, looks hideous. The smaller the area clear cut, the smaller the environmental and aesthetic impact.

Jurisdiction	Type	Maximum area
Ontario		260 hectares
Tasmania	Slopes less than 20 degrees	100 hectares
"	Slopes steeper than 20 degrees	50 hectares
British Columbia		40 hectares
Sweden		20 hectares
Latvia	Dry sandy soils	10 hectares
"	Dry soils, not sandy	5 hectares
Poland		4 hectares
Slovenia		Generally forbidden since 1949
Austria		Over 0.5 hectares with permission, maximum 2 hectares,
Switzerland		None since about 1900

Economic approach

Foresters readily accepted the idea of multiple-use management, but it was not until the 1950s that forest economists started to work on how economically rational decisions might be made if there were more uses than wood.[11] They thought that the theory of joint-production of two or more commodities that was used in agricultural and resource economics could apply equally well to the forests. They illustrated it in the simple case of rangeland forests that produced both wood and forage for cattle; if the stands were kept dense, there would be more wood and less cattle but, if they were kept open with plenty of grass for more cattle, there would be less wood. Water provided another simple case as young, rapidly growing forests used more water than old ones. With the relevant production, cost, and revenue information for wood, cattle or water, the economists could calculate the best mix. At the heart of this theory, was the belief that trading the different uses against each other in a make-believe market would enable the ideal balance to be detected.[12] Even the best choice between mutually exclusive uses, such as between wood production and wilderness, could be detected if the economic value of each was calculated.

Real cases were never as easy or as simple as their theory. First it was hard to get good information. Although they knew a lot about wood, the foresters had little quantitative information to answer questions such as about the relative effects of how many more cattle could be grazed per hectare, or how much more water could be collected, if stand density was reduced by half? Experiments could be started and measurements made to address this sort of problem, but it was harder to find the proper prices.

The economists classified the values in a forest into those that came from using it directly – like wood that was either sold in markets or consumed locally; or those that were indirect – like its value for protecting a water catchment; or those abstruse values which people even far away attached to its very existence.[13] Before they could use their theories, the economists needed to assign conceptual prices to everything that was not actually sold in real markets. Recreation was the first non-market forest use that they attempted to value; although it was the easiest, it was still difficult. The foresters could not find out exactly how many people came into most large public forests as they could come through different routes at any time or season. The best they could do was monitor popular places, count the traffic entering on main routes and ask a sample of the visitors how far they had come. The economists assumed that the visitors' costs in time and travel represented the value of the forest to them. Then, they calculated the total value to society by relating the number of visitors and their values to the population in the places they came from. The whole process was riddled with questionable assumptions and could not show what would happen to its recreation value if the forest was changed.

As the environmental era progressed, more and more people valued the preservation of various species and their habitats without necessarily having a direct connection to them. The best way economists could find of assessing this was to ask them: 'How much would you be prepared to pay to have this place preserved?'[14] It was a difficult method to use because it relied on people being informed about the case before the question was asked and it required very careful design to select the sample of people to be asked and how the question should be phrased. Although it was used quite widely, it too was riddled with questionable assumptions.

These and other economic methods of valuing forests suffered from two related limitations. First, was their blindness to equity; people varied widely in their ability to travel or their willingness to pay to preserve the environment. Second, was their failure to recognise that different social groups valued the results of change differently; workers likely to lose their livelihoods if a forest was closed for strict preservation valued it differently from people concerned about the fate of rare animals.

Rather than squeeze the multiple interests and values held about forests through a make-believe market, a method was borrowed from the operations research field which recognised that each group of people had their own criteria.[15] It too relied on people being well informed about possible changes to the way a forest was to be managed. The analysts asked people to rank the importance they gave to possible changes under environmental, economic and social criteria. For example, in an Australian forest region in which there were fiercely conflicting views, comprehensive information was gathered and five options for using the forests were prepared.[16] Representatives of thirteen different groups were asked to rank the options in the three main criteria and their sub-criteria. Their preferences were added together to show that in this case the two most preferred options were the most extreme – either high preservation or high timber production – while the three options that provided intermediate positions between the extremes were less favoured; a result that mirrored the polarised society. While the method recognised the complexity of society, it contained the questionable assumption that adding the preferences of different groups together represented the preference of society as a whole.

One way or another, through political pressure or moral suasion, the environmental issues were increasingly recognised in operations and planning practice. Although the economists had seldom been able to estimate values *before* decisions were made, they could calculate how much of the value of commodities would be lost *afterwards* if, for example, the opportunity of producing wood was reduced to allow for, say, habitat protection.[17] It was in this way that economic appraisal proved to be of most use in planning the future of the forests.

Comprehensive planning

The progress made in applying operations research methods to planning the future of industrial plantations in the southern United States was extended to the more difficult problems of the native forests. It was an urgent matter for the US Forest Service. When in 1976 the government made multiple-use planning compulsory, the foresters had to be able to demonstrate that they could undertake it in a rational and comprehensive manner.[18] Not only did they have to consider the relative values of all the renewable resources, they had to do so in a systematic and interdisciplinary manner and involve the public.

Two of the many difficulties they faced can be mentioned here. First was the large size of the American national forests. Each plan had to cover 630,000 hectares on average, and some were considerably larger.[19] Unlike the plantations, with their single species grown in even-aged stands, producing a single product for a single owner, each national forest contained several different types of forest, most with several tree species in multi-aged stands. More inventory data was needed to describe them, the models to predict their future wood yields were generally less well developed and more information was needed about their multiple uses. Although the state was the single owner, there were multiple interests, 'stakeholders', lobby groups, agencies and levels of government involved.

The second difficulty struck directly at how forestry – the art and science of managing forests – was conceived and how the foresters saw themselves. Forestry had arisen as a single field of study, but internally it was a multi-disciplinary one that trained the foresters in basic sciences of botany, zoology, geology, soil and climate; specialist subjects of silviculture, regulation and timber; and applied subjects like surveying, road construction and fire fighting. The foresters were the scientists, the managers, and the arbiters of what should be done in the forests, but in the environmental era they could not longer do it all. The more detailed understandings of botanists, zoologists, wildlife experts, soil scientists, hydrologists, biometricians, economists, computer systems analysts and others all had to be brought to bear. Planning had to be done in multi-disciplinary teams in which the foresters were no longer the sole arbiters. Indeed, when public participation was taken seriously, plans could no longer be formed in professional isolation.

The foresters hoped that their tradition of rational planning for long-term wood production could be enlarged to include the comprehensive requirements of the environmental era. To this end, the US Forest Service developed a computer modelling system and set up teams to plan the future of all its national forests.[20] Like the system that had been developed in Georgia for industrial plantations, it sought the best combination of the ways each part of the forest could be used, while meeting various obligations, or 'constraints', like providing certain quantities of timber while reserving certain areas as wildlife habitat and keeping within the budget for operations. It considered each decade for a period of 150 years, longer than was needed for the faster growing plantations. It took 'best' to be the economic benefit, or 'net present value', of whatever the economists could value, mostly timber. As so many attributes could not be valued, and the users and interests were so varied, it was used more as a simulation model to compare the effects of alternative options, than as an economic model that could provide the socially optimal solution.

The foresters' hopes that they could incorporate the environmental issues into the economic and rational planning models that they had used for wood were over-ambitious, especially for the vast American forests. Even though every discipline needed was brought into the planning teams, there was never enough data or knowledge to quantify all the aspects of the forests; nor, in the politically charged atmosphere, were plans likely to be accepted by all parties. An Australian case can illustrate this.

Victoria chose the Otways Forest as its pilot project to see if the modelling systems developed in the United States would enable its economic, environmental and social objectives to be balanced.[21] Comparatively, it was a smaller forest of 157,000 hectares that was valued for its timber, water, recreation, scenery, flora and fauna. There was good information about the yields of timber and water, which depended on the age of the stands, and some information about the relative abundance of four species that also depended

Box 11.2. Example of Comprehensive Planning, Kootenai National Forest, USA[22]

The Kootenai National Forest in Montana lies in the Rocky Mountain Ranges on the border with Canada, and is one of America's largest. Its 9,000 square kilometre area contains mountains, rivers, lakes and dams for hydro-power. One fifth of it is above the tree line, is precipitous or carries no forest and much of this and other areas have never had roads into them. It is prized by for its scenery, wilderness and recreation values. There is habitat for endangered species – grizzly bears, wolves and eagles – and also for elk, which are hunted in the forest. Slightly less than half the area carries the forest that supplies the timber industry on which local towns and communities depend. There were conflicting views about how the balance between protecting endangered species and wilderness, recreation and producing wood should be found.

Planning started in 1979 with public consultations to identify all the issues to be considered in detail in the forest plan and its environmental impact statement. Modelling proceeded, drafts were issued in 1982, further modelling and consultations occurred until the number of possible futures was whittled down to six that represent different ways of striking a balance. The final plan was issued in 1987.

Six alternative options, Kootenai National Forest, USA

Major theme of option	Net present economic value ($millions)
Present practice and limited budgets	460
Provide big game habitat, using elk as indicator species	658
All roadless areas, 1,633 square kilometres to be wilderness	1,034
Protect roadless areas, 329 square kilometres be wilderness, emphasise non-motorised recreation and visual quality	1,064
Cost efficient for timber, no wilderness	1,143
Final plan provides combination of wilderness, roadless areas, wildlife protection, recreation and timber.	733

The plan was reviewed annually and changes were made in response to new information, regulations from other agencies and public pressure, all of which have reduced the amount of timber sold in the five years to 2010 to one quarter of the amount sold at the start of the plan.

on age and could indicate faunal response.[23] There was no way of including recreation, scenery and other issues in the model so that it was impossible to balance the government's objectives analytically. The detailed geographic information system needed for the model and other quantitative information also guided the overall plan. It identified buffer and other areas where logging should not take place and showed how timber and water production could be scheduled more efficiently. It also identified the most important areas for further research. Extensive consultations with individuals, interest groups and parts of the bureaucracy gained the eventual plan credibility and local acceptance. Although the plan was adopted by the Government in 1992, it was not accepted by State and national environment groups who successfully pressured the Government in 2005 to reserve the forest as a national park from which timber production would be excluded.

In spite of their limitations and cost, the 'hi-tech' planning models provided a systematic way of examining issues and providing a base for discussion.[24] However, as cases such as the Kootenai and Otways Forest Plans showed, the foresters' hopes of including the environmental issues comprehensively within their foundation framework of rational planning could not be realised. There was an important difference between the models developed for industrial plantations and those for the natural forests. The plantation models were designed to optimise efficiency within an agreed system, but the visions of desirable futures for the forests could not be reconciled. One, often associated with the philosophy of 'deep ecology' and expressed strongly in countries like the United States, Australia and New Zealand, believed that large parts, or even all of, the native forests should be 'saved' from human use. The heart of its planning problem was not efficiency, but how to allocate large areas of land between conservation reserves and multiple-use forests, as discussed in the next chapter. The other was the foresters' vision of human use. They had to revise some of the basic concepts on which they had built forest science in light of the advances in forest ecology, but had the strand of close-to-nature forestry in Europe to draw on.

Ecological approach

The evolving field of forest ecology provided the scientific base for a new approach for understanding the complexity of the forests.[25] It arose in the midst of the controversies about how the forests should be used and it created a critique of the silvicultural foundations of forestry.[26] It informed both the attempts to manage the uses of the forests in the balanced way, discussed in this chapter, and the design of conservation reserves discussed in the next chapter.

Forest ecology both stimulated and drew on general ecological concepts. For example, the idea was advanced in the 1930s that a 'succession' of stages occurs after an area has been 'disturbed' by fire, storm or people that leads to a stable 'climax' type of forest. It could be seen readily in conifer forests dominated by a single species, but the way in which forests with several species changed was less predictable. The idea that each species might find its best place along an 'environmental gradient' provided a way of explaining their distribution through the forests according to slope, aspect, soil and drainage. The idea of an 'ecosystem' was advanced, also in the 1930s, to emphasise the interconnections between all the organisms in a forest, with each species occupying its particular environmental 'niche'. It was known that small islands could support fewer species than large ones and this idea was extended to forests. Areas reserved as habitat had to be large enough or linked through 'wildlife corridors' for species that ranged widely. Put together, these ideas meant that the ecology of a forest had to be understood on a 'landscape scale' as well as on smaller scales.[27] Influenced by these ideas, the foresters' concept of multiple-use management evolved into a wider one of 'ecosystem management'. In the twenty-first century a batch of ideas about complexity led to forests being viewed as 'complex adaptive systems' that it was argued should be managed accordingly.[28]

One of the major environmental critiques concerned clear-cutting, the simplest and most widely practised of all the silvicultural systems.[29] It involved felling all the trees in a 'coupe' at one time and regenerating the area to grow a new even-aged stand. It suited most of the conifer, and some eucalypt, forests that grew naturally in even-aged stands of a single dominant species. It was also a ready way of restoring the productivity of forests that had been badly depleted in the past. Genetically improved trees could be introduced and it could produce uniform crops to suit industry. But its simplification was the antithesis of the ecological approach that cherished the forest's complexity. Removing all the cover from the soil increased erosion and created a jarring image in the landscape.

The foresters took several ways to address the problem. Their operational approach was to restrict the size of the coupes allowed under the codes of forest practice, mentioned earlier. Another was to distribute them across the landscape and design their shape and position to give them a more natural

Figure 11.1. Clearcut area in high quality, wet sclerophyll forest, Styx Valley, Tasmania, 2004.

Photograph: June Orford.

Figure 11.2. Clearcut area in low quality, dry sclerophyll forest, TPFH Concession, East Coast, Tasmania, 1980.

Photograph: John Dargavel.

appearance. They could also be designed to suit the habitat requirements of particular species as these became better understood.

The foresters' silvicultural approach was to enlist some of the established European systems and adapt them to current ecological concerns elsewhere under a banner of 'variable retention silviculture'.[30] In its simplest form, they kept a few 'habitat' trees on the coupes. This was a useful practice in Australia where many possums, birds and bats nested in the hollows of old trees. It answered immediate needs and, if some younger trees were kept long enough to 'grow' their hollows, it would provide for such species in the future. However, the forest ecologists stressed that the whole 'structure' – from dead branches on the ground, to grasses and herbs, shrubs, small trees and the large, old trees – was needed to provide habitat for all the other animals, insects, plants and fungi.

Nature-based silviculture

The ecological ideas of the forest as a living community that had had been presented by Biolley and others in the first half of the twentieth century were developed further by Hans Leibundgut in Zurich from the 1950s. An increasing number of foresters in Central Europe accepted them during the 1960s and 1970s until they became known as the 'nature-based forestry' movement.[31] The name comes from the title of a much earlier book that advocated selection systems, rather than clear-cutting, so that the forest canopy could provide continuous cover.

The core of nature-based forestry was a vision of creating a silvicultural synthesis between ecology and economics. On one hand were the conceptions for understanding a forest biologically as a living community; while on the other hand were the physical and economic conceptions that aimed for the highest volume production and financial yield.

Rather than see them as contradictory, the nature-based forestry movement envisaged regulating the natural forest to achieve a sustained yield, while treating it as a living community. This was not only desirable, it was essential. In the environmental era community expectations of conserving the multiple values of the forests had increased and the foresters had to change their emphasis from wood production to conservation. They also had to keep the forests as productive parts of rural societies and this had become more difficult as timber prices had gradually fallen in real price terms since the 1950s, while labour costs had risen.

Box 11.3. Hans Leibundgut (1909–1993)

Hans Leibundgut

Photograph courtesy of Anton Schuler.

Hans Leibundgut was born in Neuravensburg, Germany in 1909 to Johann Leibundgut, a cheese-maker, and his wife Elise. He studied forestry at the science and technology university, *Eidgenössische Technische Hochschule*, ETH in Zurich. After gaining his diploma he worked in the Communal forest of Couvet, managed by the Swiss Control method, and in the forest of Lenk in the Lötschental valley of the Bernese Alps. In 1934 he returned to ETH for post-graduate studies as scientific assistant to Hermann Knuchel and other professors there. His 1938 doctoral thesis tackled the forest and economic situation in Lötschental. In 1937 he started to work as a forester in Büren an der Aare in the gentler Bern region, where he gained further practical experience in practical forestry. However, three years later he was appointed professor for silviculture at ETH in Zurich, a position he held until he retired in 1979.

Leibundgut developed the thesis of close to nature forestry and promoted it in Switzerland and elsewhere in Europe. He considered the forest to be a living community so that silviculture had to rely on the ecological situation and the foresters had to take into account the different phases of development of each stand. However he also looked at the economic aims. Because of the long rotation periods he wanted to avoid unnecessary investments by making an advantage of the natural processes of development in the forest. In developing his thesis he worked closely together with scientists in mycology, phytopathology, entomology, ecology and soil science.

Leibundgut had a particular interest in virgin forests, which he considered to be a guide for close to nature silviculture. He promoted the establishment of conservation reserves to safeguard rare and endangered species. After World War II Leibundgut was a consultant to FAO and the Slovenian government and from 1965 to 1969 he was the Rector of the ETH Zurich. He was the editor of the *Schweizerische Zeitschrift für Forstwesen* (Journal of Swiss Forestry) from 1946 to 1979. He was awarded doctorates by the Universities of Munich and Vienna and was an honoured citizen of the community of Wasen.

Figure 11.3. The selection system in the Regensberg communal forest, Switzerland, managed to keep an almost continuous forest cover. Note log on ground awaiting removal.

Photograph: John Dargavel.

During the 1960s, 1970s and 1980s, the leading silviculturalists from the Universities of Ljubljana, Munich, Zürich, Florence and sometimes Nancy, met more or less informally, to discuss how forests around the Alps could be managed in ways that were close to nature. They sought their inspiration from the dynamics of virgin forests dynamics. In the same period a group of foresters formed the Arbeitsgemeinschaft Naturnahe Waldwirtschaft (Working Party for Close to Nature Forestry) to support the nature-based movement. In 1989 Dušan Mlinšek, from the University of Ljubljana and past president of IUFRO, called a meeting that merged both in an association, named ProSilva.[32] It now operates in 25 countries to promote the principles of close to nature forestry.

The ProSilva movement promotes 'forest management strategies which optimise the maintenance, conservation and utilisation of forest ecosystems in such a way that the ecological and socio-economic functions are sustainable and profitable'. It was a cultural challenge, as the foresters had to consider themselves not only as producers, but also as protectors of the ecosystem as a whole. The ProSilva movement spread throughout Europe and its vision of nature-based forestry has become the main contemporary trend of Central European forestry. Synonyms are near-natural forestry or close-to-nature forestry.

In adopting the concept, the foresters had to question their former ideas about what was profitable and had to include maintaining biodiversity and all the recreational, protective, socio-economic, and cultural aspects. Their general goals were: to maintain continuous forest cover; encourage a diverse structure of age and size on an intimate scale; include natural processes of regeneration and competition; and intervene as seldom as possible. It was an all-embracing vision, but the foresters had to work out practical ways of doing it, based on the climate, soil and species of their particular forests. And economically, they had to increase income and cut costs to keep the privately-owned forests as viable units.

Although it takes decades to change the structure of a forest, French, Swiss and German foresters have been able to translate their vision into effective practice.[33] They have shown that changing to selection systems in which regeneration occurred naturally saved the cost of replanting after clear-cutting. Revenue had been increased because the growth of the most valuable species and grades, such as the hardwoods used for furniture, or the highest quality used for veneer, had been favoured. Studies showed that over twenty to thirty-year periods steady changes in the economic results had been demonstrated. They also showed that what was possible depended on the type and condition of each forest, the detailed work of skilled foresters and market opportunities. Moreover, their application took place while the technology of logging and industrial processing was demanding greater mechanisation, larger operations and more uniform wood; all making nature-based silviculture more difficult to achieve.

Foresters elsewhere started large experiments to see if some of the European 'close to nature' silvicultural systems could not only be adapted to meet the ecological requirements, but also applied on an operational scale to quite different types of forest. For example, a whole series of large experiments was started in the mid-1990s in the Pacific Northwest region of the United States and a similar one was started in Tasmania.[34] Although some early results appear promising, it will take longer before ecologically-based systems can be proved to be successful.

Sustainable forest management

The foresters faced great difficulties in balancing these evolving ecological ideas with the production and economic dimensions. Their difficulties were depicted as being due to a radical shift in the foundations of scientific forestry from one of simplification to one of complexity.[35] However true this may prove to be, forestry is a prosaic business and the foresters had to address the concerns as they arose. While all of their operations, economics, planning and their ecological approaches were problematic, and were the subjects of continuing scientific research and political pressures, they still had to run viable enterprises. In the state forests, the foresters might adopt the very best practices only to find that governments had committed more wood to industry than they could properly supply, as in some Australian cases. However, the ideals of close to nature forestry have proved economically effective for many private forest owners in Europe.

The United Nations' 1992 policy ideal of sustainable development, referred to earlier, had on one side been criticised as impossibly difficult, while on the other it had been greeted as the hope for the future. For the forests, it had been expressed as 'sustainable management', or 'ecologically sustainable management'. This too was seen as both difficult and a hope for the future. It is now regarded more as a process of continual endeavour and improvement. We return to it in Chapter 15.

Chapter 11

Notes

1. Salisch 2009, first published in 1885.
2. Introduced in Chapter 2.
3. The theory of multiple forest functions forests was developed in the 1930s and 1940s by Victor Dieterich (1944) and later expanded as the 'Kielwassertheorie'.
4. A principle for the US Forest Service set down in 1905 and thought to have been penned by the American forester Gifford Pinchot.
5. United States of America, *National Forest Management Act* 1976.
6. Government of Victoria, *Timber Industry Strategy* 1986; authors' emphasis.
7. Formally, the United Nations Conference on Environment and Development. See Humphreys 1996 and 2006 for the origins and subsequent development of these policies.
8. McDermott, Cashore and Kanowski 2010 make international comparisons across 38 jurisdictions in 20 countries.
9. Dargavel 2010.
10. There are many varieties of clear-cutting done in strips and patches, or with retained seed trees and shelterwoods: see Chapter 2.
11. Gregory's 1955 paper in the first issue of *Forest Science* marks the economic interest in the problems.
12. Pearse 1990 classifies the relationship as competing, highly conflicting, complementary, independent, constantly substitutable and mutually exclusive.
13. Kengen 1997 provides an overview and bibliography.
14. It is called 'contingent valuation'.
15. It is called 'multi-criteria analysis' or 'multi-criteria decision making'. The field of operations research was introduced in Chapter 10.
16. The series of Regional Forest Agreement studies conducted in 1995-2000 with ongoing work was the largest forest planning study ever undertaken in Australia. The trial of Multi-criteria analysis was conducted in the Southern Region of New South Wales by Wendy Proctor (2009).
17. Termed the 'opportunity cost' of lost commodity production, in this case by habitat protection. They could use such changes in the same way that they had analysed the impacts of developments to estimate how they would affect other sectors of the economy through multiplier effects: see Chapter 10.
18. US *National Forest Management Act* 1976, mentioned earlier. See Chapter 11 for the methods used to plan for industrial plantations.
19. The *average* size of a US national forest is about one sixth the size of the *total* forest area of Austria for example.
20. The FORPLAN system, Kent *et al.* 1991.
21. Dargavel *et al.* 1995.
22. USDA Forest Service 1987; Kent *et al.* 1991.
23. The four indicator species were yellow-bellied gliders, long-nosed potoroos, gang-gang cockatoos and bristlebirds.
24. Barber and Rodman 1990 provide a scathing review of FORPLAN's shortcomings, many of which are recognised in Kent *at al.* 1991.
25. See textbooks such as Kimmins' 1997 update of his 1987 textbook and Lindenmayer and Franklin 2002.

26. Puettmann, Coates and Messier 2009 provide a concise history and discussion.
27. Lindenmayer and Franklin 2002 developed the idea of a matrix of issues in different spatial and temporal scales.
28. Mitchell and Beese 2002. Puettmann, Coates and Messier 2009 describe the various theories of complex adaptive management, but their proposed approach is probably too recent for our historical perspective.
29. See Chapter 2.
30. Franklin *et al.* 1997; Mitchell and Beese 2002;
31. Leibundgut, 1949, 1989; Pockberger, 1952; Thomasius 1992; Johann 2006.
32. The founding members came from Germany, Austria, Belgium, Slovakia, France, Greece, Hungary, Norway, Switzerland, Slovenia and Croatia; most of them were well known silviculturists teaching at universities or practitioners. The first President was Brice de Turckheim from France.
33. Turckheim 2006; Wobst 2006.
34. Peterson and Anderson 2009; Baker and Read 2011.
35. See for example Ciancio and Nocentini 2000.

12.

EXCISING

Great areas of forest had to be excised from human use if their rich diversity of plants and animals – their 'biodiversity' – was to be preserved, or so a powerful argument ran. It was a different vision for the future and the antithesis of the foresters' vision of human use and of the forest science they had founded. Although in the environmental era they had sought to balance production with other uses and values, as described in the previous chapter, it had not been enough; specially protected areas were still required. This too developed forest science, as described in this chapter, but it was advanced primarily by ecologists, not by foresters.

The vision of protecting special areas of forest had origins as old as human history and it inspired the national parks and wilderness movements of the twentieth century, although different countries saw these quite differently as they tried to unravel the contradictions between excision and human use. For example, Switzerland wanted to conduct ecological research into natural processes free from human interference when it formed its first national park in 1909.[1] The area was uninhabited, but hunting, grazing and fishing were banned and the public could walk only on marked trails. By contrast, parks were created in the United Kingdom to ensure public access to areas of natural beauty, first by the non-government National Trust at the very end of the nineteenth century and then in 1949 by legislation that formally allowed public access to large areas of private land. The British parks remained inhabited and were farmed to maintain their traditional landscapes.

The vision of protecting special areas for distinct environmental objectives was gradually recognised in nature protection legislation distinct from the general forest legislation. At the start of the twentieth century, fifteen European countries had general forest legislation, but social attitudes to nature were changing.[2] Nature conservationists urged respect for the beautiful, the sublime and the scientifically valuable. Nature should be protected in the same way as monuments and art. It became a worldwide movement that we can illustrate first from Germany and Austria.

Josef John, head of the Vimperk forest district on the Schwarzenberg estate (in the now Czech Republic), was one of the first to rue the loss of what we now call cultural and ecological values. He realised that virgin forests were needed to provide the knowledge on which to base the management of other forests. Thanks to his long-term efforts and those of the provincial forester, Count Schwarzenberg decided in 1858 to keep some of his old-growth forests in the Zátoň forest district as permanent reserves called the Boubín virgin forests. His example was followed by some other forest estate owners.[3] A study of endangered natural habits led to a state conservation office being set up in 1906 and a non-governmental *Verein Naturschutzpark* [Nature Conservation Parks Society] being started 'to protect original and impressive landscapes and their natural communities of plants and animals against civilisation'.[4]

The vision of protecting nature was becoming well established in Europe by the start of the twentieth century. The Netherlands (1908), Switzerland (1909), Sweden (1909) and Norway (1910) were in the vanguard and other countries like Bulgaria, Denmark, Finland, France, Greece and Spain soon followed. Perhaps the vision was only strengthened by the devastation of World War One, because it was written into the new German constitution in 1919.

The defensive preservation of the pre-industrial forest landscape was an intimate part of the overall protective vision. For example, Romania's Slătioara Forest of old-growth was reserved in 1908 and Germany's first strict forest reserve was established in Württemberg in 1911. The alpine flora and fauna were the focus of several laws (Poland 1868, France 1913, Austria 1886 to 1920) and of a general drive that stimulated other countries to establish parks.[5] Between the World Wars, parks and nature reserves became quite popular and were established by several European countries.[6] Many of these European parks and reserves were quite small and were well accepted, or at least did not cause great social and economic disruption. Not so the American or imperial visions for nature protection on a large and brutal scale.

The American vision for protected areas was one that not only preserved areas of great scenic beauty, but saw them as wildernesses where the awe of nature could be experienced, or the frontier of exploration imagined. It was a vision of a landscape without people, without indigenes; a landscape in which people from the industrial cities could renew their spirits; it was a landscape to be visited, not lived in.[7] The first US national park was created at Yellowstone in 1872 at the end of the Settler-Indian Wars. Its indigenous Miwok people were evicted but kept sneaking back until 1969 when the Parks Service burnt their last settlement.[8]

The imperial vision, declared in 1900 at an international Conference on African Wildlife, was for game parks to be managed for European hunting safaris. It led to large areas being declared in South Africa, German East Africa (now Tanzania, Burundi and Rwanda) at the start of the twentieth century.[9] Generally, the indigenous people could remain but were prevented from hunting by game laws and in apartheid era South Africa they were largely hidden from visitors. The Society for the Preservation of the Wild Fauna of the Empire (now the Fauna and Flora Preservation Society) was formed in 1903 to recognise the widening interest in fauna preservation. It led to reserves being declared for particular species, such as those for the tiger, elephant, orang-utan and others in the Netherlands East Indies (now Indonesia) in the 1930s for example.[10]

As the conservation movement gained momentum, more parks were created and the different visions concentrated on the preserving the flora and fauna. In 1947 conservation took its formal place on the world stage – half a century after IUFRO – by forming the International Union for Nature Protection, which in 1956 changed its name to the International Union for the Conservation of Nature and Natural Resources (IUCN). It

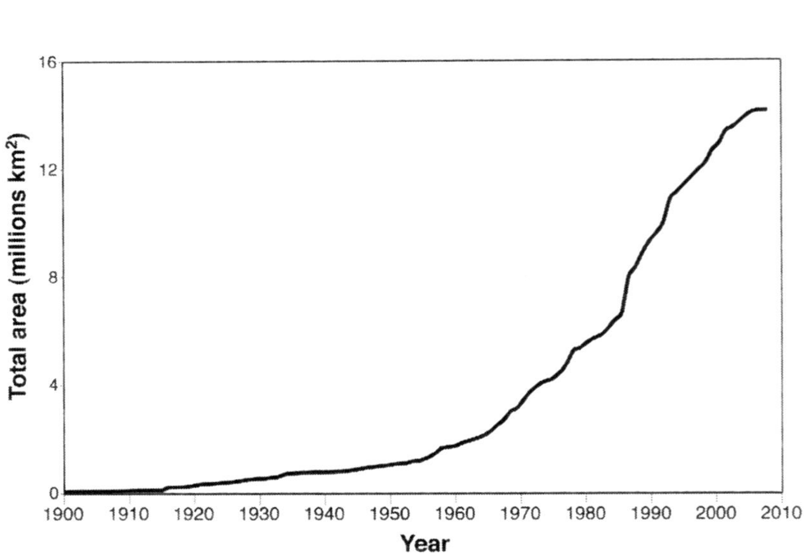

Figure 12.1. Area of nationally designated protected terrestrial areas, 1872 -2008.

Source: Drawn from IUCN and UNEP 2009. *World Database on Protected Areas.*

grew to become a large body with states, government agencies and non-governmental organisations as its members. It provided an authoritative voice for environmental science in public policy debates.

The area of specially protected parks and reserves grew rapidly during the environmental era from 2 million to 14 million square kilometres (Figure 12.1). Smaller areas of marine parks were added, mostly from the 1980s. The rapid increase was largely a response to the deforestation and degradation of many of the world's forests and the consequent risks to their biodiversity. The extent and severity of these changes had to be measured, the risks to each species assessed, and new reserves designed to best effect. These were areas in which forest science had to be advanced, but it had to be done with a sense of urgency and in a highly politicised atmosphere.

Deforestation

It was the rate at which tropical forests were being cleared that that caused most alarm. Turning forests into farms and towns had always been essential, but overzealous clearing had led to wood shortages, erosion and alarms about the drying of the landscape.[11] It had prompted the rise of modern forestry, but in the environmental era it was the concern about the loss of wildlife, even the extinction of some species, that turned the established interest in fauna protection into an urgent concern; one that expanded from its national and imperial origins to become popular worldwide.

If the alarm over deforestation was to be understood on a world scale, its extent and rate of progress had to be put on a quantitative basis. This was difficult because how much different countries knew about their forests varied greatly; some were more willing to provide information than others; and at least two measurements were needed to estimate change. Although deforestation had its general meaning as loss of forest, it had to be defined more closely if it was to be measured. This too was difficult because the definition of 'forest' itself was often confounded by its old legal meaning as a category of royal tenure, with its common meaning of a mostly tree-covered place, and with its ecological meaning as a category of vegetation. The definitions have varied between countries and the purposes for which they are used so that as many as 800 may have existed. Several are in current use by different United Nations organisations.[12]

Every five to ten years from its start FAO had collated the statistics supplied by national governments on the area of their forests.[13] It asked them to distinguish between 'closed forest', where the trees cover most of the ground, and 'open forest' or 'woodland'. It refined the definitions in successive surveys, but this made comparisons unreliable. Estimates were also made by other investigators and agencies so that by 1990 at least 26 existed for the closed forest alone. An increasing number of reliable estimates came from individual countries which the World Resources Institute assembled into influential reports that put tropical deforestation on the international political agenda in the 1980s.[14] This increased the organisational complexity of forest science and the diversity of its funding.

Measuring the forest area was the most basic step towards any quantitative understanding. The foresters had measured the European forests and some of those in other lands with laborious ground surveys, and had used aerial photography from the 1950s, but neither method could cover the whole world's forests and do so consistently time after time. The scientific and technical breakthrough came in 1972 when the United States launched the first of its 'Landsat' satellites – the seventh was launched in 1999 – that could capture images of the whole earth, piece by piece every sixteen or eighteen days.[15] Each image was taken in several bands of colour from blue-green to infra-red, mostly at a resolution which meant that objects smaller than thirty metres could not be distinguished from their surroundings. The images came back to earth in an endless stream of digital data that could be handled only in the most powerful computer centres, and by specialist analysts.

The bands of colour could be combined to make the images appear to be in natural colour, or they could be variously combined to highlight different types of vegetation or uses of the land. Forests could be picked out in a general way by anyone who knew a particular landscape, but doing so reliably was more difficult. The analysts knew that each type of forest, vegetation and land surface had its own colour 'signature' which, once determined, could be used to interpret the images automatically. It was a long process to determine even the major types and many areas were unclear. They had great problems with the forests. Although they could readily identify dense forests whose tree canopies covered the ground and could pick conifers from the lighter coloured hardwoods, not all forests were dense, so that their colour signature came from the ground, grass and shrubs as well as the trees. The open woodlands of the arid zone presented the greatest difficulty with this, because their colour signature was affected by the time the image was taken; early or late in the day the dark shadows of the trees obscured the grass. Many of these problems were addressed by improved technology in the later Landsat and other satellites and also by more sophisticated ways of processing the images.

Although the analysts could make reasonable estimates of forest areas in the United States and a few other countries by the 1980s, the ability to do so more widely depended on obtaining access to the images – politically difficult during the Cold War – and having the computer facilities and skills to analyse them. These difficulties were overcome in 1997 when reliable and consistent estimates were needed to measure the deforestation for its carbon emissions (discussed in Chapter 14).[16] However, it was not until 2010 that FAO was able to publish consistent estimates for the world, based on the Landsat images taken in 1990, 2000 and 2005.[17] It showed that the rate of deforestation was higher than previously estimated – 16 million compared to 13 million hectares a year between 1990 and 2000 – but was decreasing slightly. The regional differences were stark: South America and Africa had lost large areas of forest, as had Indonesia and other parts of Southeast Asia; whereas Europe was gaining forest as old farms were abandoned.

Better estimates of deforestation only reinforced what had been depicted since the 1980s as a crisis for the tropical rainforests. While the absolute loss of forest was the greatest issue, the degradation of these and many other forests was also of great environmental concern and added to the drive to excise forests for conservation reserves.

Degradation

Much of the drive in founding forest science had been to prevent or restore the loss of productive capacity caused by misuse, but in the environmental era it was enlarged to prevent or restore the loss of biodiversity. Although there was a general sense of what a degraded forest was, as somehow impoverished with its best trees removed and *different* from a normal healthy forest, defining it for international comparisons proved difficult.[18] Like deforestation, different definitions of degradation were used and modified by United Nations organisations.

One of the oldest distinctions was drawn between 'primary' and 'secondary' forest; the former being described as the 'primaeval', 'virgin', 'old-growth' or 'climax forest' and the latter as whatever forest followed some great disturbance to the primary forest, be it fire, storm, logging or temporary clearing. Some of the secondary forests might recover quite naturally if given enough time, aided perhaps by the foresters' tending. The ecological path of recovery could vary in many types of forest, so that the species composition of a particular patch might change somewhat. However, some forests might be so severely altered that some species would be permanently lost, making them truly degraded. By defining primary forest as areas that have no visible signs of human disturbance in the satellite images, the distinction could be used in FAO's world survey. It could quantify the obvious: South America with its Amazon forests had the greatest proportion of primary forests – three quarters – while Europe had virtually none.[19]

From a world perspective, the risks of losing biodiversity due to deforestation and degradation were greatest for the tropical forests of the developing nations. There were also risks in the temperate and northern 'boreal' forests of the developed nations where the conservation movement was strongest and where ecology was most advanced. Even though the boundaries were secure in Europe, North America and Australia, and the forests were managed to balance their uses and values, large conservation reserves were called for. Several reasons were advanced: they could retain scarce examples of unmodified forests; provide sanctuaries for rare species; be baselines against which the effects of human disturbance could be measured; and avoid unexpected risks from the cumulative effect of disturbances that independently were of little account.[20] They could also provide wilderness for jaded urbanites.

Vulnerability

At the heart of the drive to excise forest areas from human use lay the sense of how vulnerable they were to damage. The ecologists encountered a dilemma in deciding how to define and measure this; their ecological understanding encompassed all the forest's organisms, but some are more vulnerable than others. They could not examine all the species and it was urgent to understand the situation of iconic species, such as tigers, that were known to be in danger, but they also had to examine the threats to whole ecosystems.

Before they could start to gauge vulnerability, the ecologists had to be able to assess the habitats and take inventories of the species. Their history echoed the foresters' struggle to map and measure just the trees a century before, but although they had the foresters' maps of forest types to give them a start with the vegetation, they needed categories based on mixtures of all the species, 'vegetation assemblages' that did not necessarily correspond with foresters' mixtures of the trees. Botanical ecologists surveyed strip lines, 'transects' through the forests or set out grids of small plots, 'quadrats', to record the species and map their patterns. While their work was laborious, at least the plants did not fly, slither or run about and, although a few stung, none bit, poisoned or mauled them.

Wildlife ecologists had the harder time as they captured small reptiles in pits, mammals in traps and birds and bats in nets (or recognised their songs); and as they recorded animals killed on roads or searched hunters' records of bears and deer, for example. The rarest animals were hardest to estimate, but they counted paw prints, droppings and 'scats', recorded sightings or photographed nocturnal animals with surveillance cameras. Reviewing all the methods is outside the scope of this book, but the essential points were that the wildlife ecologists often had to use several methods to make their estimates, which still carried considerable margins for error. They had to determine whether a population was balanced in its ages and sexes and could breed future generations. And they had to find out what habitat each species needed for its food, security and, critically, for its breeding. The vegetation and wildlife had to be seen together.

The ecologists' prime concern was for those species at the greatest risk of extinction. They started to list them from what they knew, either nationally or for the whole world. Few lists had been published before the mid-1960s, but a rush of activity followed the publication of IUCN's first *Red Data Book* in 1966. It listed the mammals and birds at risk throughout the world and was soon followed by IUCN lists for reptiles, amphibians and flowering plants; and by numerous national and specialist lists in the 1970s.[21] The subjective judgements of the early IUCN lists were replaced in the 1990s by the publication of the supporting data, peer reviews, practical guidelines and clear definitions to distinguish between species that were 'critically endangered', 'endangered' or 'vulnerable' to extinction. The lists were extended to cover more groups of plants and animals and were revised when the status of particular species needed to be altered in the light of new information. The process of compiling the red lists involved scientists from many countries as contributors, reviewers and assessors in what it was one of the largest continuing cooperative efforts ever undertaken in science.

The red book process directed attention to saving particular species from extinction but, while it was systematic and rational, it was not *ecological*; it was not directed to whole systems. It was too late. Whole areas needed to be saved *before* they became so degraded that any of their species became vulnerable. This needed to be planned.

Box 12.1. Tiger Conservation in the Sundarbans

Tyger, tyger, burning bright.
In the forests of the night,
What immortal hand or eye
Dare frame thy fearful symmetry?

The tiger's fearsome reputation was captured in Blake's poem at the end of the eighteenth century, when more Europeans were encountering it in the wild. In Eastern cultures the tiger had ancient meanings, but the poem epitomised the 'otherness' of the Orient in the Western imagination. Tigers were then widespread across Asia from Siberia to Indonesia but are now found only in limited areas, with about half of their depleted population being in India. IUCN has them on its red list of endangered species.[22]

The Sundarbans in the delta of the Ganges is one of the tiger's remaining habitats. It is an area of changing river channels, tidal waters, low-lying islands, poor farms and mangrove forests. Its history is one of multiple and changing values.[23] A remote and inaccessible sanctuary in the pre-colonial period, it became increasingly settled and developed from the late nineteenth century. However tiger attacks hampered forest clearing and land reclamation, giving the Sundarban beasts their legendary reputation as man-eaters. When they were thought to be killing 1,600 people a year, the Indian Government declared them to be vermin in 1883 and put a bounty on their heads. Over 2,400 were killed for the bounty, while others were killed by villagers for safety or shot by Europeans for sport and trophies. The revenue potential of the area led to a large state forest reserve being declared in the 1870s with consequent restrictions on local use. A century later its potential to protect the 250 or so remaining tigers changed its purpose.

Even though they did not have good survey data, the wildlife ecologists knew that the population of tigers in Southeast Asia was falling drastically. Something had to be done. In 1969 IUCN resolved to ask governments to declare a moratorium on killing tigers until they had completed their population censuses and ecological studies.[24] India responded with legislation and in 1973 launched 'Project Tiger' by declaring the Sundarbans forests and eight others as special reserves.

The ecologists thought that each tiger needed to have ten square kilometres of forest to have enough food and that three hundred tigers were needed for a healthy population. Finding 3,000 square kilometres was impossible in the Sundarbans, where only 1,330 could be found for a core area of human exclusion, but this could be surrounded by a similarly sized area of buffers where some local use is allowed. Several hundreds of people were evicted in 'resettlement' schemes, but a few people remained in buffer areas. The World Wildlife Fund and other international NGOs supported Project Tiger in its censuses and research projects. Greater international attention was also drawn to the Sunderbans when it was recorded as a World Heritage Site in 1987 and an International Biosphere Reserve in 2001.

Periodic censuses in the Sunderbans reserve appear to show the number of tigers to have been held fairly steady as a result of these measures. New photographic survey methods may make these estimates more reliable. However, poaching to provide ingredients to be smuggled out for the Chinese medicine market remains a major threat. Internationally, the tiger remains on IUCN's red book as a species in danger of extinction.

William Blake [1794] 1991. 'The Tyger'.

In *Songs of Innocence and Experience* (Princeton, N.J., William Blake Trust/Princeton University Press).

Designing reserves

The ecologists' ideal of planning exclusive reserves in a rational way echoed an older foresters' ideal of ensuring that each country had its 'ration' of forest to support its society. But allocating forest for either sustaining yields or preserving biodiversity mixed ideals with power which might or might not draw on scientific knowledge. Although there were schemes by landowners, only nations could create legally-binding conservation reserves. During the environmental era, governments were increasingly influenced to do so by international ideals and powerful organisations, both governmental and non-governmental

Every ten years the IUCN reviewed progress in creating conservation reserves and in 1992 declared its ideal that each country should put at least ten per cent of each its ecosystems into reserves it called 'protected areas'. There were so many different types of these in parks and conservation reserves in the world, with as many as 500 definitions in existence, that its first step was to establish and define its categories. Its highest class was for areas managed for science or wilderness; its next two classes were for areas where the ecosystem was protected but recreation was allowed, or specific natural features were conserved. It then recognised areas that needed management action to conserve them or that were important landscapes. Its lowest class was for areas that were predominantly unmodified and were managed to preserve the biodiversity while allowing some use for local communities. The foresters' managed areas and plantations were outside its pale. The ideal of excising areas from current or possible human use had already taken hold in most nations (Figure 12.1) so that IUCN's target was easily met by 2000 and has now been exceeded, even though a few regions – North Africa, Oceania and Southern Asia – have not reached it.

The ecologists needed to identify the most important areas for preserving biodiversity. It was a thorny problem as the number of species, their abundance, the extent of their habitat and the threats to their continuation all had to be evaluated. The most direct way was to follow IUCN's ten per cent ideal but, if there was very little of some types left, then that would be insufficient. A base line for the calculation was needed. How this could be done varied between nations.

Australia had three quarters of its tall, commercially important forests growing on state land and a vigorous environmental movement. It had followed the international *Convention on Biological Diversity* set at the UN's 'Earth Summit' Conference in 1992 with its national strategy in 1996 and legislation. It set its base-line as the extent of each ecosystem as it had existed in 1750, well prior to the start of European settlement in 1788. The ecologists' first task was to estimate where these ecosystems had been. Some were clearly apparent, but others had to be reconstructed in geographical information systems from the existing vegetation, herbarium and museum records. Australia set its target, not at IUCN's ten per cent, but at the more ambitious fifteen per cent. As many of the forests had been cleared for agriculture and towns, some more stringent targets were set based on the area of what remained: sixty per cent of the vulnerable and old-growth forests, ninety per cent of high quality wilderness and all of the rare and endangered types. Australia already had 10 million hectares of forests in conservation reserves, a further 13 million were transferred by 2008, leaving only 9 million in multiple-use state forests where timber was produced. Although the targets were largely realised on state land, Australia has 100 million hectares of woodlands and forests that are privately managed and are outside the conservation reserve system.

The conceptual approach, historical background, size and proportion of forests dedicated to strict conservation vary widely across Europe in ways linked to the type of forest and its history. The term 'strict' reserve is interpreted very differently in different countries: in many cases controlling game and fire and removing invading exotic species are allowed. The American concept of complete non-intervention is not realistic in Europe, which needs a more versatile concept. Forest protection includes different degrees and types of restrictions on forest areas with regard to their use. Forests are selected on a regional basis to form a network of protected areas with different degrees of protection. In Finland, for example, they include national parks, strict nature reserves, wilderness areas, protected peatlands, protected old forest areas, protected lake shores, herb-rich forest protection areas, ridge protection areas and forests in Lapland that were protected to prevent the northern timber line shifting.

The European Union made great efforts to protect natural habitats and species by setting up a coherent ecological network of conservation areas under the title of 'Natura 2000'. It consists of sites nominated by the member countries and creates a legally-binding obligation to protect them. It also provides a chance to point out the importance of nature protection at the national level.[25] For example, many sites in the forest landscapes need to retain their low-intensity agricultural practices in order to conserve them, yet these are lost when farming is abandoned.[26]

Landowners in many countries have made voluntary efforts to conserve their forests.[27] Austria has a tradition of doing so that dates back to the nineteenth century and the most important remnants of undisturbed alpine forests today owe their existence to these voluntary initiatives. In this early period it was less common to establish protection areas or conservation reserves in public forests by decree. Although national parks, Natura 2000 and other areas have now been protected by decree, some landowners continue to protect their forests voluntarily. In 1995 the Austrian Forest Reserves Programme was started to support them in conserving biodiversity. The programme is based on legal conservation contracts agreed to between a forest owner and the Republic of Austria for reserves selected in natural forests by specialists in a research centre. Now there are 188 reserves covering different types of forest spread in a network across the country.[28]

Excision and balance

The two visions for the future of the forests – of either quarantining their biodiversity or managing them for human use in some balanced way – co-existed during the environmental era; they still do. Each had its passionate adherents and moral claims, with social and economic consequences which tossed them into the pits of power and politics. But each generated advances in forest science that were imbued with the hope that better knowledge would lead to better outcomes for the forests. In science many of the advances, although stimulated by one or other of the visions, applied to both.

By the end of the twentieth century, it was increasingly clear that excision alone could not preserve all of the world's biodiversity. However much was in protected areas, there would be greater areas outside them, with few exceptions, of which New Zealand is one. The best hope, so some of the ecologists argued, would be to take a landscape approach that embraced not only the large protected areas, but also the measures in the multiple use forests on both state and private land: they had to be seen together.[29]

The science of each of these visions was drawn from the biological disciplines, with applied and managerial emphases in the case of forestry. Neither was drawn from or sympathetic to the social sciences; indeed the wilderness and excision concepts denied humans a place, while forestry in its foundation and imperial forms carried a militaristic approach to control.

The ecologists and foresters not only advanced forest science, but carried the different visions into their professional practice. They knew what needed to be done. Not all agreed. Different visions for the forests were advanced, new forms of forestry were created and forest science was advanced in ways that were drawn from the social sciences, as described in the next chapter.

Notes

1. Kupper 2009.
2. Dimitz 1907.
3. Such as 1888 forest reserve Buky u Vysokého Chvojna (eastern Bohemia), 1892 Kočevje region (Slovenia), 1903 forest reserve Šerák-Keprník (Silesia), 1904 forest reserve Labský důl (eastern Bohemia), 1909 forest reserve Javořina (southern Moravia), 1910 Switzerland (virgin forest reserve at Scattlé near Brigles).
4. A memorandum by Hugh Conwentz 1904 prepared the ground for the foundation of the first office for nature conservation in Germany, which he headed when it was founded two years later.
5. The first National Park in the Western Alps was set up in Switzerland in 1914. In the Eastern Alps it came into being when an area of 4,000 ha was donated by a private forest owner to the German-Austrian Alpine Association for the purpose of establishing a national park in 1918, which became the core of the present Hohe Tauern National Park (set up in 1984). In Bulgaria in 1931 the Strandzha Mountain became the first reserve and in 1934 the Vitosha National Park became the first National Park of its kind in Bulgaria and the Balkan Peninsula.
6. e.g. Romania 1904, Sweden 1910, Spain 1917–1918, Italy 1922, Romania 1927, Portugal 1930, Poland 1932, Ireland 1936, Lithuania 1937, Greece 1938.
7. Cronon 1996 and many other writers have argued that the concept of wilderness is deeply conflicted and draws a false dichotomy between people and nature.
8. Dowie 2009. The term 'conservation refugees' is used by Dowie and others to describe indigenous and local peoples evicted to create conservation reserves.
9. Carruthers 1997, Mackenzie 1988, Jepson and Whittaker 2002, Sanseri 2009.
10. Jepson and Whittaker 2002.
11. See Williams 2003 for a magisterial history of deforestation. The history of dessicationist theories of drying the landscape and climate is covered in Grove 1995 as well as Williams. Before long-term climatic change had been understood, deforestation had been suspected of causing the Sahara desert.
12. Schoene *et al.* 2007.
13. An estimate was published by Zon and Sparhawk in America in 1923.
14. The World Resources Institute is a non-government think tank based in Washington, D.C. Two major activities occurred in 1985, when FAO prepared its *Tropical Forest Action Plan*, endorsed at the World Forestry Congress, and the International Tropical Timber Organisation was founded to link the 23 countries that exported tropical timber with the 27 countries that imported it. Both provided guidelines and support for balanced forest use without deflecting the trend to increase the specially protected areas. They were developed in collaboration with other international organisations such as the United Nations Environment Programme, IUCN, IUFRO, the World Bank

and other funding bodies and non-governmental organisations. See Humphreys 1996, 2006.

15. The eighth is scheduled to be launched in 2013.
16. Under the 1997 Kyoto Protocol any reduction in deforestation could be counted towards a country's target for reducing greenhouse gas emissions. 1990 was the base year against which deforestation was to be measured.
17. FAO 2010.
18. Schoene *et al.* 2007; Sasaki and Putz 2009.
19. FAO 2010. The assessment of primary forests needed expert interpretation and was not quite complete, as information was not available for the large forests in the Congo Basin.
20. Lindenmayer and Franklin 2002.
21. Burton 1987.
22. Chundawat 2011. There are six sub-species, all of which are endangered.
23. Chakrabarti 2009.
24. IUCN Tenth General Assembly 1969, Resolution 15.
25. Paar *et al.* 1998.
26. Parviainen *et al.* 1999.
27. For example, Bourke 2012 records the history of voluntary conservation reserves established in Australia since the 1930s.
28. The Research Centre for Forest Natural Hazards and Landscape (BFW) in Vienna is dedicated to researching and safeguarding the natural development of biological diversity.
29. Lindenmayer and Franklin 2002.

13.

DEVOLVING

Human voices have claimed places in the structures of forest power from time immemorial with varied successes, defeats and accommodations. They have never been silent or singular; nor in modern times have the dictates of foresters or ecologists, however scientifically knowledgeable, been imposed everywhere. During the environmental era, with which this section of the book is concerned, the claims came from several sources and levels, from local voices 'heard' only through their actions, to village councils, to peak international organisations. Some were about efficiency, because good forestry has always needed local knowledge and adaptation to circumstances. Some were an attempt to escape the world system in dreams of village independence. Some were driven by a libertarian belief in devolving government to the lowest level possible. Some hoped for social justice. Many carried democratic claims for women, for indigenous people or marginalised groups. Most hoped that by devolving power into local hands local actions would be more effective than centralised ones. They became recognised in policy documents, development aid projects and in forms of forest management known as 'social forestry'. There was a social richness in all this, but understanding it needed the social sciences more than the biological ones.

Paradoxically, these human claims over the forests gained official recognition at the same time as economic and political power was becoming ever more centralised; they became another strand in globalisation's net of contradictory trends and processes.[1] Much of what has been written about social forestry concerns such politics and policies, but our concern in this chapter is with the forest science that they stimulated. To include the social sciences within forest science was a radical and, to many scientists, an unwelcome change from how forestry and forest ecology had been conceived. The foresters had asserted their silvicultural and sustained yield principles and the forest ecologists had asserted their conservation principles; neither had allowed for other principles, other values, the diversity of human societies or the contrariness of human nature.

The claim for public participation in development decisions had been recognised early in the environmental era through environmental impact statement legislation and similar processes in most Western countries.[2] Over time, these processes had been adopted by international aid agencies and most other countries, at least in name. While they sometimes led to large projects being modified and occasionally abandoned to accommodate the variety of environmental and other voices, they applied only within the dominant belief that the path to development lay through modernisation and industrialisation. It was the challenge to this belief that led social forestry to diverge from the other forms.

Although social forestry presented an alternative vision of forest development 'from below' and 'for the people', it was gradually able to gain official recognition from the 1970s, aided perhaps because it had ancient roots and a widespread present reality. Although the old forms of common land and peasant management of woodland had been almost obliterated in the developed world, they could still be traced and their environmental and heritage values were revealed by historical and ecological research.[3] In the imperial world many local people had been ejected from the state forests, as they had been from many of the national parks and conservation reserves as these were being created. Outside their boundaries there were forests set aside for local use, or still in local hands. And beyond these, there were great areas of forests like the Amazon or the woodlands in Africa that were still being managed in traditional ways by people still intimate with their lands. All these situations were changing, which, with the varied claims, led to different types of responses (Box 13.1) that fall loosely under the label of 'social' or 'community' forestry'. Before turning to the responses, we need to sketch the changes 'from above' and 'from below' that stimulated new directions in forest science.

From above: international development

The United Nations promotion of industrialisation and international trade during its First Decade of Development in the 1960s had not produced the sustained economic growth in many developing countries that it had hoped for. In the forests sector, many of the projects for forest industries, plantations and timber exports that FAO and other agencies had fostered had not been successful for a variety of reasons and the benefits that Jack Westoby and others had expected had not trickled down to the poor.[4] For its Second Decade of Development in the 1970s, the United Nations promoted a more equitable distribution of income and wealth and a greater emphasis on health, social welfare and education.

FAO's Forestry Department realised that it needed to promote the place of forests in general rural development, in addition to its usual projects for producing wood from forests and plantations. This became urgent in the oil crises of 1973 and 1979, when the price of kerosene, also called paraffin and widely used for cooking in developing countries, suddenly soared. Strangely echoing one of the founding concerns of scientific forestry, wood as a source of energy gained worldwide attention. FAO's statistics in the 1980s reported that almost half the world's wood was being used for fuel, but the difference between developed and developing countries was startling: the developed countries used less than one fifth of theirs for fuel, whereas the developing countries used three quarters.[5] And anyway, who had included the few sticks a woman gathered to cook her family's meal? Or their importance?

FAO took 'Forests for People' as its theme for the Eighth World Forestry Congress in 1978. Jack Westoby, in a startling reversal from his previous analysis, captured the moment by castigating the failures of past policies – 'the famous multiplier effects are missing' – and reporting that 'very, very few of the forest industries that have been established in the underdeveloped countries have made any contribution whatever to raising the welfare of the urban and rural masses'.[6] He called on the world's foresters to 'work for the day when the rich and many-sided contribution of forestry is harnessed to the service of all'. His was an eminent but lonely voice calling for the structures of forest power and knowledge to be challenged. There were other voices, but these were in the forests.

From below: forest voices

Voices in the forests are muffled when we listen for them or when we think of them metaphorically; only occasionally are they strident. The difference in level is important, although it needed historians, anthropologists, sociologists and political scientists to point it out. They showed that the peasant revolts over forest rights, that had been the stuff of history books, had occurred against a background of the 'everyday forms of resistance … foot-dragging, dissimulations, false compliance, feigned ignorance, desertion, pilfering, smuggling, poaching, arson, slander, sabotage, surreptitious assault and murder and so on'. Those were the 'weapons of the weak' against the powerful.[7]

The foresters and ecologists in their state organisations were the powerful in their control of state forests andconservation areas and in places in their regulation of private and communal land. But their control was limited: even the boundaries of most forests were hard to determine, let alone mark; people could enter or leave at many places; and who could say where an unmarked tree or an animal had come from once it had been removed? The weak helped themselves to wood for fires, poles for their huts or litter for their animals; they nudged their farms into forests and let their animals stray to graze whenever they could. Their history *versus* the foresters can be only glimpsed in cases stretching back into the medieval and imperial courts because those were the instances that were prosecuted; most were unrecorded. Their actions were typically opportunistic, flourishing when and where control was weakest, as during the English Civil war discussed in the introduction to this book. Of course, it was not only peasants who did so; others took opportunities where they could.

In the environmental era, the scale of local people's involvement with the forests in developing countries was only gradually realised internationally. It took different forms of which three can be mentioned here. First and the least obvious were the indigenous people living in the tropical rainforests. By the end of the twentieth century, it was estimated that there were 60 million of them, and a further 400 to 500 million people directly dependent on the rainforests.[8] They were the most vulnerable to the expansion of logging or other developments. Second, and the subject of great attention, were the people who invaded the forests in increasing numbers to clear and farm their own patches, or who ventured further in from surrounding areas for their needs. Wherever new roads were built these trends accelerated so that whatever might have been intended officially, pushing logging roads further into the forests hastened their degradation. This constant nibbling away was harder to gauge than large deforestation schemes that could be seen on satellite images.[9] It made the push to excise areas into conservation reserves all the more urgent, yet, by expelling people, it only increased the pressure on other forests. Third was the expansion of population that changed the whole scale of peasant resistance. In the developed countries resistance had already withered away, but in the developing countries it flourished, producing a crisis in forest control.

It was the Chipko ('chipko' means to hug in Hindi) protest movement in the Garhwal region of the Indian Himalayas that captured the greatest attention, especially in the West where the feminist movement was evolving at much the same time as the environmental movement. Peasant protests had a long history in the region, but in 1974 they took a new turn in the forests of the remote hill village of Reni.[10] Its forests were part of its subsistence in that its ash trees provided the wood needed for the agricultural implements.

In spite of previous objections, the Forestry Department marked many of Reni's trees to be felled and sold. Choosing a time when the men had left to complain to the foresters, the contractor sent his men in to fell the trees. In a spontaneous action, the village women rushed into the forest, clung to the trees and prevented them being felled that day.

Protests spread throughout India. Some followed Ghandi's example with long marches that exerted political pressure by echoing the independence movement and casting the Forestry Department as the oppressor of the people. Meanwhile, in some parts of the country, the local pressure for firewood increased the covert thefts of wood to such an extent that the existing regime of policing the boundaries of state forests could no longer function. New forms of managing the forests had to be found.

Social and community forestry[11]

The foresters started from what they knew.[12] They could fix the firewood and desertification problems by establishing village woodlots and encouraging farmers to plant trees; it was just a question of doing things on a smaller scale, or so they thought. International aid agencies and NGO's provided funds to start projects. For the foresters it was a question of finding the right native or exotic species that would grow quickly, produce good firewood and provide poles for building huts; and then teaching the villagers how to get on with it. The foresters' view was too simple. Although the many projects had different results, two cases can illustrate the general point, one from Niger, the other from Nepal.

Niger

In 1986 Canada's International Development Research Centre reviewed the social forestry projects that it had supported in Africa since 1972.[13] Some of these were in Niger, a poor, dry, land-locked country on the southern side of the Sahara desert. Although there had been successful projects for establishing collective village woodlots in parts of Asia, the results in Niger were disappointing and a sociologist was appointed to investigate. The sociologist interviewed the villagers and found that there was virtually no communal land available for woodlots or compensation for peasants whose land was used and any woodlot they planted had to be strictly regulated by the Forestry Service they distrusted. Although the people were interested in reforestation, the community woodlot model was unworkable in that society and a socially appropriate one had to be found. The answer lay in encouraging the peasants to plant trees along the boundaries of their own land, training the women in how to raise seedlings in nurseries, and in teaching children how to plant them. The people wanted the forest law to be changed so that they would clearly own the trees they planted. Although the review found that some further conventional forestry research was needed to find better planting methods in the difficult environment, the critical input was from sociology.

The Niger case is an example of how sociology brought its own methods, formal ways of collecting *qualitative* as well as quantitative information, tests of validity and so forth – its 'epistemology' – into forest science. It did not sit easily with the quantitative, experimental epistemology of forestry, ecology and the other natural sciences. In spite of this some foresters were willing to try new forms when their foundation models proved unsuccessful, as in the Nepal case.

Nepal

The development of social forestry projects supported by Australian aid in the Middle Hills region of Nepal is often cited as one of the success stories, but it only emerged after conventional means had been tried.[14] They had been prompted by the shortage of fuel wood in the capital, Kathmandu, and by the apparent erosion on steep hillsides in many parts of the country.[15] Australia had provided advisors to the Forests Department and had aided its reforestation projects since 1962. They had tried planting fast-growing eucalypts and the native chir pine, they had provided a store to keep seed in good condition and had devised nursery practices to improve the survival of pine seedlings; all technically sensible measures. Although they had demonstrated that the river red gum, a eucalypt, could grow well in the deep soil of the valleys, that land was needed for agriculture. In the harsher Middle Hills region they had to rely on chir pine and, in spite of their work, the plantations were not successful; the fences were continually breached, goats ate the young trees, the trees were replanted the next year, only to have it happen again and again; even where the trees did survive, the local population used the plantations to gather fodder. By 1975 the Australian foresters realised that their conventional model of establishing government plantations could never succeed against the local population; there was not enough suitable land and it was needed for local subsistence.

International influences could be seen in 1976 when Nepal's national forest plan included a section on public participation and allowing people to grow trees for their own benefit. Two years later, FAO published an international guide for developing community forestry projects that repeatedly stressed the importance of understanding the social and cultural norms of communities.[16] It noted that the traditional training of foresters did not leave them 'well equipped to deal with people', but clearly there were exceptions, as when a few Australian and Nepalese foresters, such as T.B.H. Mahat, developed a new approach for an area in the Middle Hills region.

The new approach had three elements. First, it was based in those communities whose village councils, *panchayats*, wanted to take part. Although they had to operate within the general forest law, they decided where and what should be done. More panchayats took part as they learnt of the early examples. Second, the Australian aid project provided the support and training for local people to grow chir pine, fruit trees and other broad-leaved species in small nurseries. The resulting plantations were necessarily small, with an average area of about thirty hectares for each panchayat, and were established on what was formally state forest land, although it was commonly used by local people in practice. Young broad-leaved trees sprang up unexpectedly and grass flourished within the protection of the pine plantations so that they produced fodder as well as wood. Third, the model depended on the level of personal understanding and trust between the foresters and the local people. The success of this and other community based projects led the Government to initiate similar projects. By 1995 there were so many groups involved that they could form a Federation of Community Forest Users that grew to have over 11,000 groups as members.

The Nepal case shows how the epistemology of social science found a way into forest practice. It did so partly in the foresters' pragmatic tradition of finding solutions to particular circumstances and partly through the diffusion of ideas. An anthropologist was part of the Australian project for only a brief period, but material such as the FAO guide was available to the foresters. The project also attracted and supported foresters to undertake postgraduate courses at the Australian National University. Some analysed how the benefits of the project accrued differently according to status, caste and gender; others found that they could use their pragmatic approach within sociology's method of action research.

The forestry that proved successful within the contained communities of the mountainous Middle Hills region was arguably less successful in

Nepal's more accessible and politically fractious Terai region with its more divided population.[17] Both the Nepal and Niger cases showed that the form of forestry had to suit the social, economic and political conditions if it was to be successful.

Forms of social forestry

As official agencies and non-governmental organisations directed more of their attention and funds to social and community forestry projects from the late 1980s, they became more aware of the diversity of social conditions around the world. Some recruited social scientists to help them develop socially-oriented forms of forestry; some hoped that new forms could take a share of the established structure of forest power; others hoped for independence. Many new forms were tried and given different names (Box 13.1).[18] They reflected the different types of relationship – or 'property regimes' – between the state and the forest users. Some of the most important kept the formal tenure of the forest land in state hands, but provided rights to local users. These could vary significantly, as between those in Nepal for example, where the local people kept all the benefits flowing from the forests and plantations under their management, and those in India, where the benefits were shared with the Forest Departments. The diverse arrangements in the political and legal arenas are mostly beyond the scope of this book, but joint forest management, indigenous use and non-timber forest products need to be mentioned.

Joint forest management[19]

In parts of India, the foresters had little choice but to find new ways: however many guards they had, they could not keep the people out of the state forests. Desperate for fuel, and without rights, the peasants degraded the forests, treating them as *de facto* open-access resources. National social and development policies were also changing towards alleviating poverty and in 1989 a central policy of managing the forests jointly between the States' forest services and the local people was declared. The various States developed different schemes for doing this: some paid local people to guard the forests; others provided them with shares in the products or the revenue; all were set up by agreements between the local councils and the forest services in what has been called 'panchayat raj' – rule by village council.

Box 13.1. Terminology for social and community forestry[20]

Collaborative forest management

Collaborative forestry is a broad term for partnerships between local communities and Forest Departments over the management of forest on state land. It includes joint forest management and overlaps participatory forestry.

Community forestry

Community forestry is widely used by international agencies to describe people-based forms of forestry. It includes indigenous forms of forestry, government-initiated programmes like the community user group forestry in Nepal and the joint forest management schemes in India.

Farm forestry

Farm forestry is a term to cover small-scale schemes that help farmers to grow and harvest trees on their own land, often by providing cheap seedlings and advice.

Indigenous forest management

Indigenous forest management refers to the way indigenous peoples manage forests independently of, and traditionally isolated from, the mainstream economy and society.

Joint forest management

Joint forest management is a term to cover sharing products, responsibilities, control, and decision-making over a forest between local user groups and Forests Departments. Contracts between the groups and the Departments specify the distribution of authority, responsibility and benefits. The primary purpose of joint forest management schemes for the Departments was to improve the condition and productivity of the forests. A secondary purpose for them was to promote a more equitable distribution of the products.

Participatory forestry

Participatory forestry is an umbrella term coined to cover forms such as community user group forestry in Nepal and joint forest management in India.

Rural development forestry

Rural development forestry is a term to cover the growth and management of trees where the primary decisions are made by individual or group users and where the primary benefit remains within the household or community.

Social forestry

Social forestry has been a widely used term. In the 1970s it was used in India for forestry on village land, but not on state forest reserve land. Westoby used it to mean any forestry for local needs and it was used to include individual farm forestry, communal village planting and sometimes forest management by villagers. In India for example, it came to refer to plantations sponsored by the Forests Department, with varying degrees of local participation, on village grazing commons, some government land or the sides of roads, canals and water storage reservoirs or 'tanks'.

These schemes, with their multiple objectives and different interests, needed sociologists and economists as well as foresters to examine their operation and evaluate their effectiveness. As more studies from different regions and countries were made, the degree of community cohesion was found to be a critical factor. Although many schemes did reduce forest degradation and improve productivity, the economic gains were largely captured by village elites in ways that could worsen the poverty of some groups. By contrast, indigenous and ethnically homogenous societies were found to distribute the benefits in a more equitable manner.

Indigenous and forest-dependent people

In lands of new settlement, like Australia, New Zealand or North and South America, indigenous people made up the whole population prior to colonisation, but in many Asian countries like India – and also in the far north and some other parts of Europe – they had long been ethnically and culturally distinct minorities. Some were dependent on the bounty of the forests for their subsistence and on its isolation for their separation from mainstream society. Their dependence, especially on the tropical rainforests, gained international attention as the frontiers of exploitation and deforestation extended. Initially the plight of people in the Amazon and Borneo, for example, was highlighted as part of campaigns to protect the forest environments, but later as part of other campaigns to defend their human rights. Internationally, the difference in motivation was reflected in the dates they were recognised and the legal status they obtained. For example, the legally-binding *Convention on International Trade in Endangered Species of Wild Fauna and Flora* came into force in 1975 at the beginning of the environmental era, but the United Nations did not adopt its non-binding *Declaration on the Rights of Indigenous Peoples* until 2007.

Imperial foresters and anthropologists had noted how indigenous and other forest-dependent people used the forests, but in the environmental era many formal social studies were undertaken. A theme that ran through these was the way communities allocated rights to the forests and their produce. The studies were necessarily specific to particular communities and forests, but revealed many instances of enduring patterns of indigenous knowledge, use and management – forms of forestry – that had modified but not destroyed the forests on which they relied. This was of particular interest to resource economists as it ran counter to the classical political economy arguments for privatisation that had been revised as an environmental 'tragedy of the commons'. What emerged was a better understanding of the complexity and diversity of traditional social arrangements for using resources and their vulnerability to the pressures of modernisation.[21]

Although wood production was their first interest, the imperial foresters had kept a sharp eye out for anything else that they might be able to boost on world markets. For example, Lane Poole in his surveys of Papua and New Guinea in the 1920s reported, under the heading of 'minor forest products', on the possible use of various species to provide bark for tanning, fibre for ropes and netting, rattans for furniture, resins, gums and oils, dyes, garden plants, medicinal plants and food.[22] He only occasionally noted purely local uses, such as the fragrant leaves of a shrub that young men put round their arms and were 'said to be much appreciated by members of the opposite sex'.

As the concepts of social forestry took hold, it was realised that what were 'minor' products on a world market might be of major importance in a subsistence economy, or might still present opportunities for local communities. Researchers pursued three main strands of scientific inquiry to develop ideas about how communities might be assisted, or their resources exploited. Ethnobotany, the study of plants used by different groups, had ancient roots in medicine and advanced in the twentieth century once botanists and anthropologists collaborated in their research. Attempts were made to recover some benefit from pharmaceutical companies who developed new drugs using the indigenous knowledge of their use. The second strand of research was into understanding and improving agroforestry, the systems for producing farm and tree crops together. It also had ancient origins and was widely practised. Research needed the skills of agronomists, arboriculturalists, sociologists and others. It was coordinated and advanced for developing countries through the World Agroforestry Centre, started as the International Council for Research in Agroforestry in 1978. The third strand of research investigated how all the products of the forest, other than wood, were used in developing countries. Their use also had ancient origins. In the social forestry sense they were examined for their potential to improve the livelihoods of forest dwelling and village people, rather than only for their export potential. What had been called 'minor forest products' were renamed as 'non-timber forest products'. Implicit in this was some expanded use which raised all the questions of how development, conservation, equity and local culture might interact. Research was coordinated and advanced through FAO's Forestry Department and the Centre for International Forestry Research established in 1993.

Several governments and NGOs declared policies and funded projects to support social forestry with technical advice and established institutions to share information and provide training.[23] The Regional Community Forestry Training Center for Asia and the Pacific, for example, was formed in 1987 by FAO, the Swiss Government and Thailand's Kasetsart University. It linked its activities to those of numerous other organisations which together gradually raised the effectiveness and prominence of community forestry. They argued that unless the needs and skills of local people were harnessed to alleviating their poverty, the forests could not be sustained by the existing form of state forest management.

Community forestry schemes vary widely across societies and the types of forests. A survey of successful projects in South-east Asia and the Pacific used seventeen attributes to describe their success. Some were concerns about research, education, soil and water conservation, science and technology and profitability that were typical of any form of forestry; but most were about social attributes like leadership, innovation and entrepreneurship, equitable distribution of revenue, equitable representation of minorities, democratic decision-making, consultative management and political endorsement.[24] Many of the schemes demonstrated that, by protecting the forests from degradation, they conserved their biodiversity to greater or lesser extent; others were designed as 'integrated conservation and development projects' with that in mind.

People in biodiversity reserves

The same argument – that unless local people had schemes to alleviate their poverty the forests could not be sustained – was directed to the exclusive biodiversity reserves. The ecologists had persuaded governments that the exclusive reserves were needed to save threatened wildlife from people. The subsequent eviction of indigenous and local people had created the flood of 'conservation refugees' in developing countries mentioned in the previous chapter. Alternative, community-based policies were proposed on grounds of social justice, of the ineffectiveness of policing and in cases of the need for human intervention to preserve biodiversity. The situations were complex and studying them required the social sciences as well as the biological ones. Three cases can illustrate the point.

The first comes from an 'ecodevelopment' project in the Nagarahole National Park in South India that was one of Project Tiger's system of reserves.[25] It was designed on the principles of integrated conservation and development that included consultation with the local people, government agencies, non-governmental organisations and other 'stakeholders'. These ranged from forest-dwelling tribal people to the World Wildlife Fund and the World Bank. There was no doubt about the intent – the tribal people were to be evicted and resettled on the park boundary – but all the stakeholders were to be consulted in the design of the project. Not only were there fundamental conflicts of interest between the environmentalists, the foresters and the tribal people, there were differences between agencies and between the layers within them. Researching the way this complex situation played out over several years required the various 'actors' and the 'network' of their interactions to be identified in the social system, in a strangely similar way to that used by ecologists to examine the flows of nutrients, water and energy in a forest.

The second case also concerns people and tigers; in this case in Nepal's Chitwan National Park that adjoins a wildlife reserve and one of India's Tiger Reserves across the border to create one of the most important regions for tiger conservation.[26] Local people collected forest products inside the park and in the multiple-use forests surrounding it; tourists visited the park; the army patrolled it to prevent poachers and illegal loggers; and tigers and their prey animals roamed inside and outside the park. The research was designed to test the belief that threatened species and people could live closely together. The researchers counted the numbers of people, tigers and their prey animals (mainly deer), inside and outside the park for two years. Although the period was too short for definitive conclusions that would overturn the prevailing belief in excluding people, the researchers could show that tigers continued to survive in spite of people in the forests and had adapted their behaviour when there were more people.

The third case is drawn from the indigenous Aboriginal use of the Australian woodlands. In the current climatic period of the last 15,000 years, Aboriginal people occupied the entire continent and managed its woodlands and forests with fire.[27] Notably they created small-scale patterns of vegetation and the open grassy, park-like landscapes that were a boon to the British settlers who dispossessed them. Fire had been part of the Australian environment for millions of years and the Australian vegetation evolved to adapt to it. However, as ecologists, fire scientists and historians conducted more research into the severity, frequency, type and season of burning – the 'fire regimes' – they came to realise that the vegetation had undergone some further adaptations to the *specific* ways that the Aboriginal people had used fire. That is to say that Australian biodiversity had in part been *created* by human action. The land managers found that they needed to reproduce the Aboriginal fire regimes in order to recreate the small-scale mosaics of vegetation patches that some threatened species needed in the modified landscapes of present-day Australia.

Developed countries

The ideas and science of social forestry that we have discussed so far in this chapter evolved about problems in developing countries. The developed countries were, and saw themselves, as the donors of aid and the providers of expertise, but they too experienced the paradoxes of the world's political economy and were not immune to social dislocations in their own forests. In a reverse flow of ideas, they sometimes looked to the concepts of social forestry, devised for different contexts, to tackle their own problems.

There were many parallels. For example, the experiences of indigenous peoples and minorities in Europe, or other developed regions, echoed the experiences of similar people elsewhere. Some parallels were historical. For example, rural peoples in Europe once depended on the forests for a wide range of products for their livelihoods, as many people in developing countries do today. However, the social transformation and urbanisation of the last 300 years in Europe have de-linked communities from forests, and forestry from agriculture, to a greater extent than is so far the case in developing countries. These days Europeans no longer depend on the forests and the traditional place of forests in people's lives has been replaced by the amenity and recreational needs of the largely urbanised population.

The need to revitalise the local communities in forest regions, discussed in Chapter 11, led to community forestry projects being started in many parts of Europe that were, to some extent, similar to social forestry projects in developing countries. Although they were designed to reflect the new set of values, they also sought opportunities of using timber and non-timber forest products in innovative ways. The discourse around 'community' has evolved considerably in the sustainable natural resources literature over the last thirty years, and has become more politicised. There are different *communities of interest* in forest resources in Europe with different demands, which are often overlapping and sometimes incompatible, and which can stretch beyond the boundaries of *local* communities. For example, conflicts can arise between indigenous peoples and small forest owners; between hunting associations and conservation groups, public forest agencies and rural groups, workers and industry and so on.

The maze of rural development, agricultural and forestry policies, strategies, plans and subsidies that has been created in the European Union and its member states in the new millennium is well beyond the scope of this book. It will be enough to note that they emphasise the need for active involvement and participation of local communities, self-help rather than reliance on external action, and recommend a 'bottom up approach'.[28] No doubt forest and social scientists will investigate the extent to which these ideals affect the lives of people in the forest regions.

Fourth leg

Gradually, governments and their forest departments came to accept social forestry as a valid form. One estimate of this was published in 2002.[29] It was based on a survey of the ownership of forest land in 24 countries that together had 93 per cent of the world's forests and showed that community and indigenous groups were managing over one-fifth of the forests in developing countries.

By the end of the twentieth century, social forestry had become the fourth leg in the use and management of forests. Like the formally planned management of natural forests, the development of intensively-managed industrial plantations and the exclusive biodiversity reserves discussed in earlier chapters, community forestry had its own hopes and vision for the future, its own areas of forest science, its own institutions and its own place in the curricula for training young foresters. It gave the foresters, ecologists and development workers the fourth leg to a structure on which to base their endeavours for a new millennium.

Notes

1. Dargavel 2011.
2. The US *National Environmental Policy Act of 1969* was a turning point in recognising public participation in formal planning processes.
3. Rackham 1976 produced the first of a series of pioneering works in historical woodland ecology of Britain. Johann *et al.* 2012 gives a comprehensive European overview.
4. Westoby 1962, as discussed in Chapter 11.

5. FAO published *Yearbooks of Forest Products* and now makes the data available on its web site.
6. Westoby 1987.
7. The term coined by James Scott 1985, 1989 captured the essence of previous academic work.
8. Cited by White and Martin 2002.
9. Cattle ranching schemes deforested large areas of the Amazon rainforests as palm oil plantation schemes did in Southeast Asian rainforests.
10. Guha 1989.
11. Arnold 2001 provides an overview of the development.
12. Hobley 1996 provides a chronology of phases in the development of the new forms of social and community forestry.
13. Lessard 1986. The review formed part of the UN's Brundtland Inquiry that led to the publication of *Our Common Future* in 1987 and the World Commission on Environment and Development 'Earth Summit' Conference in 1992.
14. Griffin 1988, Gilmore and Fisher 1991, Hobley 1996.
15. Although deforestation was widely held to be the cause of widespread erosion, it was later found that natural processes were the major cause.
16. FAO 1978. This guide proved to be probably the most widely read of FAO's Forestry Papers, being reprinted six times over the next twenty years and now published on the Internet.
17. Pravat 2010.
18. Arnold 2001.
19. Chakrabarti *et al.* 2005 provide a history and literature review in the first half of this paper.
20. Based on the terminology used in Hobley 1996 and *Encyclopedia of Forest Sciences* 2004.
21. Arnold 1998.
22. His survey is described in Dargavel 2006. His detailed report was published in the Australian *Parliamentary Papers* 1925, No. 73.
23. Now The Center for People and Forests. 102 donor, partner and sponsor organisations are listed in its *Annual Report* for 2010.
24. Brown *et al.* 2005.
25. Mahanty 2002. Project Tiger and the case of the Sundarbans Reserve are described in Chapter 12.
26. Carter *et al.* 2012.
27. There have been people in Australia for at least 40,000 years, during which time the climate has changed. Gammage 2011 provides a historical account of the extent to which burning practices changed the landscape and its biota. Bradstock *et al.* 2012 provide a scientific management perspective.
28. European Communities, 2003. The sustainable forest management and the multifunctional role of forests have been defined in the Ministerial Conferences on the Protection of Forests in Europe of Helsinki (1993), Lisbon (1998), Vienna (2004), Warszawa (2007) and Bergen (2009). On 15 December 1998 the European Council adopted a Resolution on a Forestry Strategy for the EU that fixed overall principles for action addressing the conservation, protection and restoration of forests. Its principles of multifunctionality and sustainability are reflected in the rural development policy of the EU by bringing together economic, social and environmental objectives into

a coherent package of voluntary measures and thus giving added value to the implementation of forest programmes of the Member States in their regions. The integrated rural development approach puts great emphasis on linkages with other policy areas and land uses, as well as on the consideration of specific socioeconomic and ecological factors. It focusses attention on the need to combine different interests and to achieve economic, social and environmental objectives in a coherent way and to take care of the regional diversity.

29. White and Martin 2002.

MILLENNIUM

The date changed and the world paused for reflection: anxiously, seriously, but also with renewed hope and purpose. War, poverty, hunger and disease were as real as they had always been, but now humans knew they were altering the climate of the whole globe. Never before had an environmental change been so universal, so enveloping, so insidious, so inexorable, so obdurate against correction. It was hard to believe. And the great hopes of a half century now seemed naive: the United Nations could no more secure peace than the League of Nations had; the Decades of Development had not prevented millions living in poverty; and its environmental programmes had not prevented degradation on a world scale.

Not all was dismay; serious reflection included achievements. Some were spectacular: jet aircraft, the moon landing, the internet. Some were boons: smallpox had been eradicated, polio almost so and life expectancy had been increased in almost every country. But most were partial and still progressing with hope and purpose. Seldom acclaimed, the United Nations and its agencies had survived the half century to be accepted as the arena in which great changes might be proposed, fought over, sometimes agreed and sometimes made. For the new millennium, it reaffirmed its purpose, declared multiple intents and set eight goals to be achieved by 2015, including environmental sustainability and eradicating extreme poverty.[1]

The United Nations drew on its agencies and on 1,360 scientists to assess the conservation and sustainable use of the world's ecosystems. Their findings for the forests were damning.[2] Although the forests supported one billion people living in extreme poverty, and had half of the world's species of plants and animals, deforestation and their degradation continued. These were not the failures of science or vision, but of governance. Most developing countries had policies for sustainable management and social forestry; many had not implemented them. To all these problems were added those posed by the changing climate.

Changing the climate would affect – perhaps already was affecting – virtually every aspect of every forest in some way; while how the forests were being managed was also affecting how the climate changed. To start to understand this scientifically required a great deal of further research, as described in the following chapter. It had to be undertaken while the models of how the climate was likely to change were themselves evolving.

Compared to its foundations a century and a half before, forest science had become more complex, less assured, in the new millennium.

Dismay over the state of the world and its forests was not confined to developing countries or to climate; it pervaded every corner. All the forces of modernisation and globalisation seem to have gone awry. Millennium declarations and goals were ostensible intents but, for anything to be done, the detail had to be worked out country by country, place by place. There were so many interactions between the economy, the environment and society that it was harder to consider the sectors, like forests, independently. The boundaries of forest science were less clear and the way it constructed its knowledge was more questioned. The forest scientists focused their investigations on the ability of particular activities to be sustained – their 'sustainability'. As they always had been, they were sustained by their hope that better knowledge would help the vision of a sustainable world to be achieved. Their endeavours are considered in Chapter 15.

In the final section of this book, we reflect on the long history of forest science, consider some of its consequences and speculate about its path as the millennium progresses.

Notes

1. The *United Nations Millennium Declaration* was adopted by the General Assembly in 2000 and set eight *Millennium Declaration Goals* to be achieved by 2015. The goals were to: eradicate extreme poverty and hunger; achieve universal primary education; promote gender equality and empower women; reduce child mortality; combat HIV/AIDS, malaria and other diseases; ensure environmental sustainability; and develop a global partnership for development.
2. Millennium Ecosystem Assessment 2005a, b.

14.

WARMING

The foresters' questions about how the forests affected the climate, and how the climate affected the forests, were familiar; they had studied them for a century, but never before with such intensity. They had used the 'desiccationist' argument, that deforestation would dry the climate, to advance their cause of conserving forests in the imperial era; and they had related trees to their climates when introducing them to similar climates, 'homoclimes' in other lands.[1] They knew that the world's climate and vegetation changed over geologically-long time scales, but to find that the climate of the first century of the new millennium would be different from that of the previous one was unexpected; that it was due to human action was shocking.

The realisation came gradually.[2] A scientific understanding of how the atmosphere controlled the earth's temperature had been built up during the nineteenth century until, at its end, the importance of carbon dioxide and the fact that it passed between the zone of living things, 'the biosphere', and the atmosphere was understood.[3] In 1896 Svante Arrhenius in Sweden was able to theorise how much the temperature would rise if carbon dioxide concentration in the atmosphere increased. However, it took half a century more before it could be measured accurately enough to show that it was steadily rising. It had been known since the 1930s that burning fossil fuels – coal, oil, gas, peat – might be responsible, but in the 1960s it became possible to show that only about half of what had been emitted had accumulated in the atmosphere: the rest must have been absorbed in the biosphere, but where? It had to be accounted for. It was not only carbon dioxide that was responsible for what became known as the 'greenhouse effect', methane and several other gases were involved.[4]

Such an enormous problem could only be tackled by scientists in several disciplines and many countries and it only became feasible for them to do so when satellites could gather some of the data and when they had computers powerful enough to process their very large models of how the climatic system operated, and how carbon and other gasses circulated between the atmosphere and the biosphere. The scientists estimated how the concentration of carbon dioxide, the temperature in different regions and the level of the sea might rise according to how much the emissions increased. How emissions might be controlled and who might do so dominated environmental politics. As the scientists developed their models in the twenty-first century, they started to estimate how the weather patterns, the rainfall and the likelihood of storms might change.

Several thousand scientists in many different institutions were employed in developing and improving the models. They had set up international committees to help them cooperate in their work. This led to a formal structure, the Intergovernmental Panel on Climate Change (IPCC), being set up in 1988 within the UN system to provide an authoritative basis for international policy. It produced its first report in 1990 in time for the UN's Environment and Development 'Earth Summit' Conference in 1992.[5] The fraught political negotiations about reducing emissions at this and the subsequent international conferences (Kyoto 1997, Copenhagen 2009, Durban in 2011) are outside the scope of this book. At Kyoto, most industrialised countries, but not the USA, agreed to reduce their emissions of greenhouse gasses down to target levels and virtually all countries undertook to measure them.

The IPCC's reports and the international conferences stimulated a surge in scientific research around the world. For the forest sector two great scientific questions had to be investigated: what was the place of forests in circulating carbon through the atmosphere, and how would the changing climate affect their productivity and their biodiversity? Answering them altered the way forest science was conducted. As we showed earlier in this book, it had been founded as a distinct applied discipline that had enlarged, and had been challenged by, the emergent conservation biology discipline during the environmental era. Such distinctions were over-ridden when many disciplines were needed to address the problems of climate change.

Carbon

Carbon was the most pressing problem. How much was there in the forests? How much of it was emitted into the atmosphere? How much captured from it? The foresters had to assess their stock, gains and losses. Only when they had done so could they re-examine their silviculture, economics and management for the new millennium. It was the measurement problem all over again, but they could start from their long tradition of measuring forests and plantations for their volume of wood and had satellite images to help them they know the area of even the most remote forests.[6]

They had already made some estimates of the total weight, or 'biomass' of trees and stands. In the 1960s and 1970s the drive to raise forest yields through breeding and fertilisation discussed in earlier chapters, had also looked at raising the proportion of each tree that could be used. In 1964 a keen advocate of this, Harold Young in the University of Maine, had excavated a sample of eight species, weighed their roots, stems, branches, bark and leaves, dried them and calculated equations from which the wet or dry weight of each part and the whole tree could be estimated from its diameter and height. With such equations, foresters could survey the forests for biomass, just as they had for volume. Young also analysed their nutrients and argued that such studies were essential to understand the complete productivity of natural ecosystems and the way nutrients were cycled within them.[7] Forest ecologists agreed and used similar methods to investigate how energy from the sun, nutrients and water passed between the trees, soil and atmosphere. They also measured the shrubs, herbs and grasses, the dead leaves, twigs and bark that made up the litter on the ground and the organic matter in the soil. Such biomass studies are laborious and expensive, but by the 1980s they had made them for 1,200 forest stands in 46 different countries. It was a straightforward laboratory process to find that carbon amounted to roughly half the weight of wood and slightly less in the leaves and litter.

These early studies provided information about particular *stands*, but information was needed about the amount of carbon at *national* and *global* levels. Providing it became politically pressing when countries needed to report their emissions against their targets, and economically pressing when schemes were started to trade emissions of carbon. The European Union (then 25 countries) started the first trading scheme in 2005, followed by other jurisdictions and voluntary 'green energy' schemes. Many of these relied on the idea that emissions of carbon in one place could be 'offset' either by capturing it elsewhere, or by reducing emissions from some other activity. Establishing plantations was a way of capturing carbon in a way that could be readily calculated for sale, but their time was limited because they ceased capturing it once they were mature.[8]

A United Nations scheme for Reducing Emissions from Deforestation and Forest Degradation in Developing Countries (REDD) was launched in 2007 to direct some of the funds expected to be generated from carbon trading in the developed industrial countries to the developing countries to compensate them for conserving their forests to reduce their emissions. Environmental organisations saw it as way of creating exclusive reserves to conserve biodiversity; the foresters saw it as a way to support sustainable forest management; and others saw it as a way of supporting local communities. With these added objectives, it became known as the REDD+ scheme. It has been criticised on grounds of equity, governance and the difficulties of measuring changes. Although some aid and other funds started REDD+ projects, it will be many years before the overall scheme can be evaluated.

The difficulties of measuring forest carbon occurred at all levels. Internationally, the forest scientists contributed what information they had to the teams building the models of the global climate. They had more information for the European and North American forests than for tropical and other forests. They had to assume that average values from the stands that they had studied could be applied to whole forests, at least until they could assess them systematically. For example, they commonly assumed that carbon made up fifty per cent of the weight of wood, yet further investigations showed that it could vary between species, parts of the tree and geographic regions from 46 to 55 per cent.[9] Overall they were likely to overestimate the amount of carbon by three to five per cent from this assumption alone. The accuracy of their estimates became more important economically when carbon was traded or was part of REDD+ schemes.

Nationally, countries had to develop their own models. Australian scientists, for example, developed a large model for the continent that estimated the carbon emissions and stock according to how the land and forests were managed.[10] It used satellite images to measure deforestation and other changes to land use and it added information about the productivity and management of the native forests and plantations. As in some other countries, the model was accompanied by a 'tool box' that enabled people outside the construction team to use it. Work continued on improving the accuracy of the model's components. For example, in the absence of local measurements, the IPCC proposed a default value of ten years for the time that roots could be assumed to decay completely after trees had been felled, yet the time Australian trees took varied by species: some took more than 85 years.[11] These small examples provide a glimpse of the difficulties of assessing the forests' place in the circulation of carbon. In spite of them, the scientists had to make estimates and improve them when they gathered better information.

At the global level, the IPCC issued a special report on forests at the start of the new millennium and updated its global assessment reports every six years.[12] These made the importance of the world's forests and savannahs become clearer: they had well over half – about sixty per cent – of the stock of carbon on land and they were responsible for two thirds of the flows, or 'fluxes' of carbon between the plants and the atmosphere. The fluxes were two-way; the plants captured carbon through photosynthesis as they built their cells, but as their leaves, branches, stems and roots eventually decayed, they released it back to the atmosphere as respiration. The delicate balance between them was an important component in the IPCC's climate models: they had to be measured.

The foresters knew that the soil and its humus were important but, until the biomass studies revealed it, few had realised that it contained as much as or more carbon than the trees (Figure 14.1). It was not only the

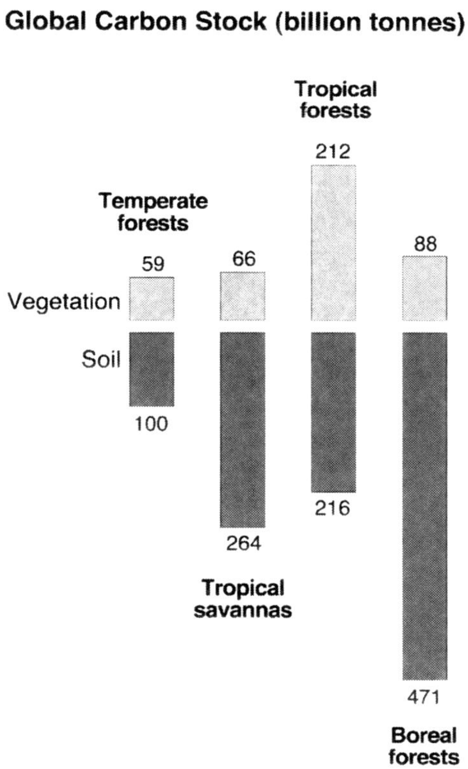

Figure 14.1. Stock of carbon in the world's forests, 2000.

Source: Adapted from Watson *et al.* 2000. *Land Use, Land-Use Change, and Forestry: IPCC Special Report of the Intergovernmental Panel on Climate Change.*

litter and roots from the trees and other plants that made up the carbon in the soil, it was also the humus and a host of microscopic and larger organisms – amoeba, bacteria, mycorrhyza and other fungi, insects and worms, for example – that break down the residues from the plants and release their nutrients to the plants and their carbon to the atmosphere. The rate at which they do so and emit carbon varies widely. It can be so fast in the tropical rainforests, with their lush vegetation and hot climates, that the nutrients can be recycled in six months, but in the cold boreal forests that stretch across the northern hemisphere up to the arctic circle it can take 350 years or more. The rate also varies according to the season, type of soil, topography, rainfall, how the forest is managed and when fires occur.

The realisation that the forests had to be managed not only for their productive values, recreation, biodiversity and local communities, but also for their carbon increased the difficulty and cost of providing scientific information. Research projects were started, and monitoring schemes started or expanded. European forest scientists were concerned about the effects of air pollution stretching across the continent and collaborated in projects to assess forest health. For example, from the 1980s they assessed the condition of the trees in a network of 6,000 plots and in the 1990s added soil and foliage assessments to estimate how much carbon the forests were capturing.

These and many other studies continued to deepen the scientific understanding of the carbon cycle and the complexity of the impacts on the forests. It was not only the changes to the temperature; there were changes to the rain or snowfall, the pattern of the seasons, and the frequency of storms that had to be considered. Moreover, as the modelling became more and more advanced, the scientists increasingly realised that the changes were likely have significant differences between regions that would affect the productivity and the biodiversity of the forests. They needed to evaluate the impacts of the forecast changes on what they understood of the complexity of the forests. Their work produced a mass of detail that will become more detailed as the millennium progresses. Only a generalised summary can be presented here (Box 14.1).

Box 14.1. Generalised Summary of Impacts of Climate Change[13]

Productivity

Gains may come from CO_2 enrichment effect, warming in cold climates and increased rainfall in some regions.

Drought

Increases in droughts are predicted for some regions (e.g. Amazon and Europe) due to rising temperature and lower rainfall. These may reduce a forest's resilience and may lead to death.

Seasonal change

Trees may be subject to greater damage in Spring and Autumn.

Pests and diseases

Rising temperatures may make boreal forests more liable to damage by defoliating insects and Norway spruce forests more liable to damage associated with bark beetles.

Fire

The risk of fire is expected to increase in fire-prone regions, such as Australia, Indonesia and Alaska. Fire in regions without previous fire regimes is expected to help pioneering trees and shrubs extend into the tundra.

Forest expansion

The dynamic global vegetation models forecast a considerable expansion of broadleaf and mixed forests into the tundra, although this is likely to be moderated by processes not yet included in the models. Other models have indicated that there may be major changes between forested and non-forested land.

Migration

It is uncertain whether forests will be able to migrate in step with the pace of climate change and it is unclear how much their migration will be impeded by fragmentation and obstacles.

Forest gains and losses

Deforestation is the major process, particularly in the Amazon and South-east Asia. Degradation is particularly important in Siberia. Area losses are only partially offset by gains in areas of plantations and the reversion of agricultural land to forests in most industrialised countries.

Biodiversity

The major factors of deforestation and degradation that threaten biodiversity will be exacerbated by climate change. Some of the areas richest in biodiversity and the montane cloud forests are especially vulnerable. The rate of extinctions is expected to increase.

Productivity

Some of the changes to the climate forecast by the IPCC were likely to damage the forests and reduce their productivity. For example, Norway spruce that had been managed and planted so successfully for timber in Europe was likely to become much more susceptible to diseases and pests; indeed, increasing damage by the spruce bark beetle is already apparent. In contrast to the damaging impacts, the warmer temperatures and a higher concentration of carbon dioxide in the atmosphere were likely to make the forests grow faster; not only would they be more productive economically, they would absorb more carbon, or so the forest scientists hoped. Their questions were how to measure it, how to predict it and how to see if it would endure.

Their first measurement to detect a long-term trend came from the common knowledge that deciduous trees burst into leaf earlier if the spring is warm. The idea of recording such seasonal cycles of plants and animals had a scientific lineage back to the eighteenth century and became known as phenology.[14] In 1957 the International Phenological Gardens project was started in Europe to provide a controlled basis for observing long-term trends. It removed genetic variation by planting cloned trees and shrubs in its series of clear sites spread across Europe and had scientists observe them with a standard protocol.[15] They showed that the trees were starting to bear leaves earlier in a way that could also be seen in satellite images; at the start of the new millennium they were a week earlier than they had been forty years before. As the trees were growing for longer they would be more productive, but in places might be more vulnerable to damage.

The effects of climate differ by latitude, altitude and the type of forest. Any change in the timing of the seasons changes the amount and date of the rain or snow that falls and consequently changes the incidence of floods or droughts. In the Northern Hemisphere longer summers delay the time when the top shoots of the trees harden and make them more liable to be killed by frost in winter. Longer summers mean that leaves of deciduous trees unfold earlier in spring and drop later in autumn; both of which can increase breakage when snow falls. Deep frost followed by warm weather over a longer period without the protecting cover of snow can kill conifers when they can not get enough water from the frozen soil. Wind damage, often followed by bark beetle attacks, may also increase when the timing of the seasons changes.

Foresters and botanists knew that tree and other species had adapted their range, 'migrated', naturally over the eons as they responded to particular temperature and moisture conditions as the climate had changed. They had investigated the climatic range of tree species when they had introduced them to similar climates in other lands (Chapter 5) and they worried whether the species in the natural forests would be able to migrate successfully in response to a faster change in the new millennium. Because trees are so long-lived, they knew that they might not notice the effects for many years, but feared that some species that now grow only in some northern and mountainous areas might become quite rare at the southern and lower end of their range by the end of this century.

The forest scientists thought that growth might also be boosted by having more carbon dioxide in the air, as well as by the nitrogen from air pollution. Given that the IPCC models forecast that carbon dioxide emissions would continue to rise, how could the forest scientists predict how much the trees might capture? They had to simulate what the atmosphere would be like well into the future. At first they tried adding carbon dioxide in greenhouses where they grew seedlings. The seedlings grew faster, but what would be the effect in more natural situations? They tried planting the seedlings in open-top chambers out in the forest into which they pumped the carbon dioxide. The trees also grew faster, but the experiment only worked until they were four years old and grew above the chambers.[16] What would the effect be in a forest? In the late 1990s forest scientists in the United States and Italy tried pumping carbon dioxide through towers set amidst three existing stands of trees up to twenty metres tall.[17] They measured the responses daily, but will have to wait many years before they can judge the effect on mature trees. In the meantime they have predicted that, although forests will continue to grow faster and capture more carbon, the rates will be less than they first thought from their greenhouse and chamber experiments. These tower experiments are so expensive that they have not yet been replicated across different sites or in tropical forests.

Estimates of the fluxes of carbon dioxide, water vapour and energy between the forests and the atmosphere were needed for the global climate models. Meteorologists initiated the global 'Fluxnet' system of over 500 recording stations to do this from the late 1990s. They had to make continuous measurements as the fluxes varied by time of day, weather, and season. They put 239 of their stations in forests but they had to build tall towers to keep their instruments well above the canopies.[18] What they found meant that some of the assumptions used in the early models had to be reconsidered.[19] It was reasonable to expect that a warming climate would hasten

Box 14.2. Tumbarumba Flux Measurement Station[20]

One of Australia's 28 stations for measuring the way carbon dioxide, water and energy flow between forests and the atmosphere was set up in the hills at Tumbarumba in Bago State Forest, about eighty kilometres south-west of Australia's capital, Canberra in 2000. Its tower rises for thirty metres above the forty metre high canopy of a eucalypt forest.

It is one of Australia's best-equipped towers. A crane holds a box of instruments clear of the tower to keep them free from eddies due to the tower itself. The box holds an infra-red gas analyser that measures the concentrations of carbon dioxide and water vapour; an ultra-sonic anemometer-thermometer that measures the direction, speed and temperature of the wind; a pyranometer that measures the radiation from the sun; and a sensor, shielded from the radiation, that measures the temperature and humidity of the air. The instruments to measure the carbon dioxide are duplicated at eight levels up the tower, and the rainfall is measured both above the canopy and also on the ground. Sixteen probes were put in the soil around the tower to measure the amount of water in the soil. The measurements are made continuously and radioed in every hour to the Canberra laboratory.

Occasionally when an instrument stopped working, the scientists filled in short gaps in the data with estimates from a nearby meteorological station. Once they had 'cleaned up' the data like this, they could use their models to see how the fluxes altered as the weather and time changed. Also, by having their instruments at different heights up the tower, they could see how the carbon dioxide, the temperature and the water vapour changed under the trees as well as above them. The scientists were intrigued to find how stable the pattern was under the trees at night, compared to how turbulent it was during the day.

Although their hi-tech instruments let them measure the atmosphere, they also needed to find out how much carbon the forest was capturing. For this, they set out an inventory of plots; some distributed across the whole 50,000 hectare forest, others relevant to the tower kept within one kilometre of it. To measure the trees, they used the simple dendrometers – tapes fastened to the trees – that foresters had used for a century to measure their changing diameters. Although they could estimate the trees' biomass and carbon content above and below the ground from these, they also had to find out how much carbon there was in the litter of leaves, twigs and bark falling from the trees, the under-storey plants and the soil. They used a well-known method of collecting the litter in traps, but measuring the fluxes in the soil was harder to do in the forest. For this, they devised traps that could absorb the carbon dioxide in small dishes of soda lime set on the soil, but it was a laborious process as each dish had to be processed in the laboratory.

When the scientists had made their measurements for a few years, they started to see how the carbon fluxes changed from year to year. When there was a drought the trees grew more slowly and captured less carbon than usual, though not as markedly as might have been expected because the eucalypts are deep-rooted trees, well adapted to Australian droughts. However, in severe droughts, the alpine ash and mountain white gum trees at Tumbarumba are particularly liable to attacks of psyllid insects that suck the sap from their leaves, killing as much as half the foliage. In these circumstances the forest, rather than capturing carbon, was emitting it.

The Tumbarumba study, like many others in the Fluxnet system is gradually revealing how intricate the interactions are between the climate and the forest.

Figure 14.2. Tumbarumba measurement station showing tower and instrument array.

Photographs: Eva van Gorsel.

decomposition and emit carbon, yet the foresters had found that some of the European forests were accumulating more carbon than they were losing. From these and other observations they thought that the forests and savannahs were absorbing about 1.3 billion tonnes of carbon more than they were emitting.[21] The climate models had taken this into account, but as results started to come in from the European stations it seemed that the effects might diminish in future.

The carbon dioxide emitting towers and Fluxnet projects were only some of the ways by which forest scientists were trying to find a quantitative understanding of how the forests affected the climate. Equally puzzling was trying to assess the likely effects of the changing climate on biodiversity.

Biodiversity

The foresters, botanists and ecologists had a good base to start from. Ever since Darwin and Wallace, they had been intrigued by the pattern of how plants and animals were distributed across the world – their biogeography.

And after they had accounted for the evolutionary differences due to the separation of continents or the isolation of islands, they knew that rainfall and temperature were the major factors, with soil, slope, aspect and drainage creating the local variations; with of course all the changes from human use. They had mapped the forest patterns at scales from major ecological zones – boreal, temperate, sub-tropical, and tropical – to single species. To relate them to climate, they had mapped the homoclimes, in some of which they allowed for minimum temperatures, monsoons and droughts. Such work had guided tree introductions, but much of it had been done graphically; for the new millennium it had to be digital.

They were alarmed about what might happen to biodiversity as the climate changed, but found that forecasting it was a highly complex problem.[22] Their first approach was to correlate the species to the climate. The meteorologists had converted their tables of rainfall and temperature to make the climate models and the foresters knew where the main tree species were from their assessment records. The botanists and zoologists had fewer systematic surveys to help them and had to search the herbarium and museum records to find where specimens, particularly of the rare species, had once been collected. They drew the data together in geographic information systems and prepared diagrams to show the 'bioclimatic envelopes' of temperature and rainfall in which particular species flourished naturally. Then they mapped where the species might be able to live if the climate was 1.5 or 2 or 3 or more degrees warmer: some would not survive.

This approach had several problems. When they first modelled it across continents they could not consider the local variations of topography. However these created variations in the micro-climate that gave many species greater resilience than they thought.[23] More fundamentally, where species occurred naturally did not necessarily define the climate to which some could adapt. For example, radiata pine occurs naturally in a restricted zone of California and Mexico, but has flourished in wider climates elsewhere. Moreover, considering species singly ignored the ecological setting on which many depended.

The 'bioclimatic envelopes' of forests had clearly moved across the landscapes in the slow pace of geological time; they had inched back to northern Europe after the ice age, for example, but it was uncertain if they could migrate fast enough for climate change. The problem was greater in the Southern Hemisphere where the ocean barred the migration of the southernmost species. Some estimates were made of migration, but there were greater problems. Many landscapes had been so altered by farming that species already sheltering in remnant patches might never escape. The possibility of more frequent fires was another threat.

The scientists developed other approaches. In one, they built 'dynamic global vegetation models' based on the fundamental cycles of energy, biomass, carbon, nutrients and water in different types of vegetation. With this they could simulate how the cycles might change as the climate changed, and also examine the possible effects of fire. In another approach, they built 'earth system models' with dynamic feedbacks between the vegetation and the atmosphere, but these sorts of models did not include how the human use of the land might also change with its consequent effects on biodiversity.

Management

The substantial scientific endeavour to examine the effects of a changing climate on productivity and biodiversity had revealed the high level of complexity and uncertainty about what might be expected. The forest and ecological reserve managers needed to work out how to apply what had been found out to reduce or 'mitigate' the changes and to adapt to them.

The foresters had some obvious ways to reduce forest emissions by halting deforestation or retaining mature forests from felling. Less obvious were the effects of different silvicultural regimes – ages of felling, intensities of felling, fertilisation and so forth. Less obvious still were the effects of reducing forest degradation by better management with codes of forest practices and similar means. They could also offset industrial emissions elsewhere by establishing plantations on already cleared land but, if funds became available from the carbon offset schemes, they could pursue more complex objectives. One example of this came from India, once the heartland of imperial forestry.[24] The intense demands for wood from the increasing population and the extent of its poverty had resulted in forty per cent of the forests being so degraded that the trees covered less than forty per cent of the ground, with correspondingly low levels of biomass and carbon stock. Restoring them to a more productive state would capture carbon – even as much as 7 billion tonnes over a 150 year period according to one estimate. It could provide social as well as biodiversity and economic benefits if implemented, at least in part through social forestry schemes.

The foresters also had to start considering how they might adapt their management to suit the changes. The most obvious opportunities were with plantations. Fortunately, the tree breeders had kept a broad genetic pool in their orchards and could start breeding strains to suit changed climate conditions. For example, in 2011 the cooperative tree improvement programmes in the southern United States started a large programme aimed at reducing the use of energy, nitrogen fertiliser and water by ten per cent and increasing carbon sequestration by fifteen per cent over the following twenty years. Building on their previous work, they aimed to produce seedlings that would be more resilient to climatic changes.[25]

The ecologists wanted to protect biodiversity but had no direct steps, like planting trees that they could take; rather they advocated strengthening or reforming existing policies. One example of this came from Australia where a strong system of forest conservation was already in place, but where there were also environmental problems and looming threats to the endemic flora and fauna.[26] The ecologists expressed the need for change, but only in the most general terms: Australians had to transform their thinking about biodiversity, more public discussions had to be held, more exclusive reserves had to be added, expenditure on conservation had to be increased and new forms of governance devised. In tropical countries, they hoped that the REDD+ schemes might eventually provide funds through carbon credits to pay local people not to log their forests.

Changing the earth's climate for the new millennium was the most startling consequence of all the processes of industrialisation, population increase and development over the previous half-century; it was the ultimate globalisation. Unravelling its complexity and interactions with the forests has only just begun and few of the actions that will be needed have been commenced. These are not all in the domains of science and international policies; much has to be understood and achieved at the local level if the forests are to be managed in a sustainable way. This is the topic of the next chapter.

Notes

1. Discussed in Chapter 8. Grove 1995 traces the dessicationist arguments back to the eighteenth century. Introductions are discussed in Chapter 6.
2. Bolin 2007 provides a history of the scientific understanding of climate change and its rise as an issue in the international political and scientific agenda.
3. Carbon dioxide is denoted as CO_2 by chemists to show that each molecule contains one atom of carbon and two of oxygen.

4. The Protocol agreed at the Kyoto Conference covered the four greenhouse gases – carbon dioxide, methane, nitrous oxide, sulphur hexafluoride – and the fluorocarbon gases that affect the atmosphere's ozone layer.
5. The IPCC produced its major assessment reports in 1990, 1995, 2001 and 2007. The fifth is expected in 2013 and 2014. Each major assessment consists of the reports of three working groups (such as Fischlin *et al.* 2007 and Parry *et al* 2007) and a synthesis report. It also produces interim and special reports, such as Watson *et al.* 2000.
6. Measurement is discussed in Chapter 2 and satellite imagery in Chapter 13.
7. Young advocated what he called the 'Complete Tree Concept' in which every part of the tree should be used. His 1964 paper with Carpenter is one of several he published with colleagues and research students. He presented the Complete Tree Concept to the American Society of Foresters in 1964 and the World Forestry Congress in 1966.
8. Strictly speaking a stand continues to capture carbon when it is biologically mature, but decay releases a similar amount so that the *net* amount it captures is zero. The stand is then 'carbon-neutral'.
9. Examples are given by Lamlom and Savidge 2003 for a temperate forest species in North America, Ma Lin *et al.* 2011 for a temperate forest in China, and Martin and Thomas 2011 for a tropical forest in Panama.
10. Waterworth and Richards 2008.
11. Ximenes and Gardner 2006.
12. See Watson *et al.* 2000 for the IPCC special report on forests and Innes 2004 for a discussion of the forest carbon cycle.
13. Derived from Fischlin *et al.* 2007.
14. An English naturalist, Robert Marsham started observing the first signs of spring in 27 plants and animals in 1736. His observations were continued by his descendants until 1958.
15. Jackson and Grace 2004. 66 gardens had been established by 1978 and 89 by 2010.
16. These are called free-air CO_2 (FACE) experiments. Jackson and Grace 2004.
17. Hendry *et al* 1999; Ainsworth and Long 2005. The tower FACE experiments were located in stands of loblolly pine, sweetgum and aspen in the USA and a poplar plantation in Italy. They were also started for grasslands and agricultural crops.
18. Regional networks are linked in the overall FLUXNET Project.
19. Grace and Rayment 2000.
20. Based on Keith *et al.* 2009; Keith *et al.* 2012; Leuning *et al.* 2005.
21. Grace and Rayment 2000 cite 2 billion tones as the amount for all terrestrial vegetation and cite forests as responsible for 66% of carbon production.
22. The 2007 IPCC global evaluation is given by Fischlin *et al.* 2007. For a national overview of Australia, see Steffen *et al.* 2009.
23. Willis and Bhagwat 2009 provide an overview.
24. Singh 2008.
25. North Carolina State University Cooperative Tree Breeding Program 2011. 55th *Annual Report.** The programmes are described in Chapter 9.
26. Steffen *et al.* 2009.

15.

SUSTAINING

The consequences of changing the Earth's climate prompted the scientific research described in the previous chapter. It was truly global in its reach and scale, but it was locally that the consequences would be felt, that action would have to be taken, that knowledge was needed. Grand policies would never be enough. The consequences would pervade the forests and their social world; their landscapes would change. Not only would the foresters and ecologists have to take action, so too would the peasants and the managers, the landholders and the companies. It was another layer of complexity that had to be included in a vision for the forest's future in the new millennium. It added to the ever-growing economic, political and population pressures on the forests.

The vision of sustainable development, launched at the 1992 United Nations Earth Summit Conference in Rio, had increasingly, if grudgingly, been recognised at the national and local levels. For this, the quality of being able to sustain the forest's future, its 'sustainability' had to be assessed on its social dimensions as well as on its economic and environmental ones. Production had to continue, biodiversity had to be preserved and ways had to be found to sustain not only all the other uses and values of the forests, but also the livelihoods and culture of people dependent on the forests. Many of the issues have been raised in previous chapters, but tackling them not only acquired a new urgency in the new millennium, it became harder. What had started as the foresters' clear management principle of sustained yield in the nineteenth century, and had expanded as multiple-use management in the twentieth century, became 'sustainable forest management' in the new millennium. It was a new vision of hope for the future, but quite what it was and how it might be achieved posed even more difficult problems for the forest scientists.

The problems they faced were occurring when there was already a much wider questioning of how scientific knowledge was created and how it depended on who was creating it. Although the great bulk of their detailed investigations could proceed much as they always had done by the normal scientific processes of classification, measurement, experiment, analysis and model-building, the larger questions unsettled the foundations of their work. In this chapter, we introduce these briefly before turning to three other areas of questioning over indigenous and traditional knowledge, landscapes and economics. All this questioning reflected the difficulty of trying to grapple with the concept of sustainability in a changing world.

Questioning scientific knowledge

The forest scientists found the questioning of how they came by their knowledge unexpected. Ever since the Enlightenment, roughly the time span of this book, they had gone about their business methodically and as objectively as they could. Of course there had been passionate disagreements, but the rules of how they should be settled by evidence and rational debate were uncontested, enabling them to advance their knowledge in the cumulative way that philosophers term 'positivism'.

One strand in the intense philosophic debates about the nature of knowledge came from historians and sociologists. In 1962 Thomas Kuhn introduced his concept of a paradigm to cover 'the entire constellation of beliefs, values and techniques and so on shared by members of a [particular scientific] community'.[1] He argued that, normally, research proceeds into the questions that the community thinks are important and, because its members share values, they can work efficiently without the constant need to question first principles. The course of forest science from its foundation through to the environmental era readily fits Kuhn's description. The forest scientists were part of the larger community of foresters who were closely linked by similar education, professional associations, scientific journals, IUFRO and the World Forestry Congresses.

Kuhn argued that substantial changes or revolutions occurred when a new set of beliefs and values arose and were gathered by a community of scientists to constitute a new paradigm. The new challenged the first principles of the old not only in the realm of ideas, but also in the social realm with all its institutions and professional identities. Although the two paradigms might go about their normal business for lengthy periods, their arguments about first principles had to be resolved eventually. Kuhn's argument was controversial at the time because it depicted science not as proceeding in a steady progress towards a greater understanding of the natural world, but as being disrupted by occasional revolutionary phases characterised by turmoil, struggles for institutional power between rival communities and personal angst. During the environmental era, the existing paradigm of forest science was challenged by a new paradigm of forest biodiversity preservation on one side and by a paradigm of social forestry on the other. Each was held by its own professional community; each brought new disciplines to forest science, which was no longer owned exclusively by foresters.

Some critiques of the established forestry paradigm were based largely on its simplification of the forest, its reduction of biodiversity and its restricted options for management.[2] The problem of how to deal with the complexity of the forest had been central to modern forestry since its foundations, but sustained yield was no longer the dominant goal; rather it was one among a number that had to be juggled in planning. We argue that in the new millennium, and in spite of the continuing political and ideological rifts over the creation of exclusive reserves, the foresters absorbed the ecological thrust of the biodiversity paradigm and attempted to adapt their multiple-use practices in the natural forests. However, they absorbed the social forestry paradigm less readily. The forest scientists continued their research into silvicultural systems in the 'harmony with nature' tradition and the various aspects of managing habitat, soils, water, recreation and so forth, but needed to give much greater attention to the social dimensions if sustainable forest management was to be put into local practice. For example, in 2007, a congress of European forest scientists reported that only ten per cent of their research was directed to socio-economic analysis, even though it was needed in two thirds of the issues of managing the forests sustainably.[3]

Questioning indigenous and traditional knowledge

Another set of critiques of modern science came from its general failure to acknowledge the value of traditional indigenous and local knowledge. There were several influences behind this. One was an increasing uneasiness about how modern science knew what it knew, its 'ontology'. Without delving into the enigmatic field of post-modern philosophy, we can note that it recognised different forms of knowledge and cast doubt on the validity of judging between them. In this sense it undermined the superiority assumed by modern Western scientists. This was particularly marked in developing countries: their forest inhabitants and the scientists knew a great deal about the forests, but they knew it differently. In the Amazon, for example, it was only from roughly the 1980s that Western scientists really started to 'see' the way the indigenous people had used and managed the forests, or the long-distance trading routes they had used to exchange products for centuries.[4]

A second influence was the increasing voice that indigenous peoples asserted nationally and internationally. For example, their rights and knowledge were recognised in 1992 in the *Statement of Forest Principles* and the *Convention on Biological Diversity* that emerged from the 1992 Earth Summit Conference; and in the 2007 United Nations *Declaration on the Rights of Indigenous Peoples.*[5] A third influence, closely related to the others, was an increasing respect, at least in some quarters, for the about 5,000 indigenous and traditional cultures, many of which continue to be threatened by the pressures of globalisation. In response, some non-governmental organisations, for example, worked to maintain languages in danger of being lost.

A fourth influence appears to have originated in the 1990s from despair at deforestation, forest degradation and the loss of the old-growth stands in managed forests, such as those in the north-west of North America. If modern science was producing such results, perhaps we needed to seek wiser ways. One source might be found in the ways pre-modern societies had managed their environments for very long periods without excessive degradation.[6] They might be the only examples of sustainable management we have, or pessimistically will ever have, but it was an ambiguous argument. On one side it was liable to be romanticised and overstated; on the other side it was unclear whether the practices of the past were relevant to the present.[7] Nevertheless, there were two strong arguments. The first was that many of today's forests and woodlands still reflect the manner by which they were shaped by traditional knowledge for long periods before the scientific forestry described in this book. The second was that indigenous communities manage about eleven per cent of the world's forests, either independently or in collaboration with governmental land management agencies.[8] Moreover, there are also many local communities who may not be officially classified as indigenous, but who also have relationships to the forests that are deeply embedded in their culture, identity and often religious beliefs.

These forest issues warranted serious investigation. For example, IUFRO focussed its 2005 World Congress on the theme of linking tradition and technology and followed up with detailed surveys and research in one of the few efforts to bridge the gap between forest scientists and the holders and users of traditional knowledge.[9] It found that, although the systems of using and managing the forests were diverse, there were common threads in communities' concern for sustainability, the relationship between people, the identification of people with their land and forests and the limits placed on exploitation for external markets.[10]

Questioning exclusive reserves

The environmentalists' policy of evicting people from the forest in order to preserve its biodiversity (Chapter 12) was increasingly questioned. Statistically, indigenous and forest-related people live in or near some 80–85 per cent of the world's protected areas and their biodiversity. Politically, the reaction to their dispossession from their homes and livelihoods caused some of the American and European NGOs to change their public stance.[11] Scientifically, detailed traditional knowledge was increasingly realised as valid, particularly by the pharmaceutical industry. Economically, the right of indigenous people to benefit when their traditional knowledge was used in this way was asserted. Internationally, ethnobiologists and linguists stressed that linguistic diversity was an essential part of indigenous knowledge and needed to be conserved.[12] Such indigenous interests were acknowledged by environmental organisations. For example, in 1996 The World Wide Fund for Nature (WWF) undertook to work in partnership with native people to conserve biodiversity on their lands; and the world's peak conservation body, the International Union for the Conservation of Nature and Natural Resources (IUCN) called for policies to support indigenous knowledge of biodiversity and respect for linguistic diversity. It was from this emphasis on conserving languages threatened with extinction that the concept of 'biocultural diversity' emerged to draw parallels between living cultures and living species.[13]

Some correlation was obvious: areas of high cultural diversity, such as in the Amazon or Papua New Guinean rainforests, for example, were also areas of high biological diversity. In the new millennium scientists started to prepare quantitative indicators that could correlate and map cultural and biological diversity together.[14] Their first global index used national level data for the numbers of languages, religions, ethnic groups, birds and mammals, all adjusted for population and area to give an equal representation to culture and biology.[15] They hoped that it would give a 'more integrated view of the patterns that characterise life on Earth'. Whether such patterns were partially caused by evolution or long practice, or were merely phenomena of remoteness are questions for further research, but they contributed to the knowledge needed to manage the forests sustainably.

There were more practical concerns as well. In developing countries, policing the boundaries of exclusive reserves was expensive and often ineffective against the increasing pressures from surrounding people. Only in rich developed countries, often with static or even declining populations, could reserves be governed easily. The focus on declaring them to conserve biodiversity did not allow sufficiently for the areas around them. They too had conservation value, especially for species that ranged far, like tigers or eagles, or for species that did not occur elsewhere. The surrounding areas were also potential sources of weeds, agricultural chemicals and feral animals, as well as grazing animals, illegal hunting and logging and all the forms of human use the reserves were designed to exclude.

In the new millennium this questioning is ongoing, especially in the tropics, where the biodiversity paradigm is splitting between strict preservation by exclusion on one side and adopting some form of social accommodation on the other.[16] The biological and social scientists need to research the issues if a sustainable path is to be found for these complex landscapes.

Questioning the landscape

Landscapes reflect the way the social and biological communities have interacted in the past, sustainably or not. They provide a sense of place for their inhabitants and can acquire a wider cultural value; some are prized, like monuments for their heritage; others valued for how they have supported communities; some, it is felt have to be conserved for the future. Although these attributes are universal, the historic and cultural values of forest landscapes have so far received the greatest recognition in Europe, where they have been included officially in the criteria for evaluating sustainable forest management.[17]

Europe's alpine landscapes have not only been highly prized for the scenic values that underpin the tourist industry, but also because their historical development provided an example of sustainable development. The idea of sustainable management can be seen in the legislation governing the use of the commons in the Austrian alpine forests and pastures since the thirteenth century, for example. The commoners were responsible for managing the woodland, the waters and the village infrastructure, with the general aim of maintaining the resource for future generations. Rules, based on the annual consumption of a farmstead, governed fuel wood and timber harvesting, the collection of litter and forest grazing by cattle, sheep and goats. The amounts of these depended on the size of the commons and the number of commoners with rights. Studies in the Bavarian and Austrian Alps demonstrated that traditional mountain farming, with its varying intensities of land use, led to a mosaic of natural or close to natural elements and supported the cultural identity of rural society.

Although alpine mountain farming had been sustained for generations, it was as vulnerable to economic change and the gradual drift from rural to urban living as elsewhere in the world. It led to an increase in the area of European forests, as, naturally or by reforestation, trees took over unused pastures; and to policies that tried to ameliorate the trend in the hope that the European alpine area could be developed in a sustainable and participatory way.[18] Such questioning of the landscape provides another way of considering how social and biological communities can interact in a sustainable manner and how the material, economic and social forces endlessly shape it.

Questioning economics

Rather than look for sustainable forest management in the traditional knowledge of indigenous and local communities, or in the landscapes of the past, the economists questioned their own knowledge. Just as the foundations of science had been questioned; so too were the foundations of economics. The mainstream 'neo-classical' economists' had long based their paradigm on the assumption that society was composed of rational, greedy, impatient individuals whose myriad independent actions in the market produced an equilibrium to society's advantage. On this they had built often analytically elegant models such as Faustmann's forest rotation model (Chapter 3) or Leontif's input/output model of a national economy (Chapter 10). From the start of the environmental era some economists advanced alternative paradigms – behavioural economics, social choice theory, institutional economics and ecological economics – based on widening the neo-classical assumptions and, in the case of ecological economics, by placing the human economy within the natural world. It was not that people were not rational, greedy and impatient, but that they were *also* passionate, benevolent and patient; and they did not act *only* as individuals in a market, but *also* acted collectively.

The economists needed these widened assumptions to see how the concept of sustainable development might be applied. Its definition as 'development that meets the needs of the present without compromising the ability of future generations to meet their own needs' raised a host of problems, especially when the needs of the poor were to be given overriding priority.[19] How were needs to be assessed? How was equity to be considered? How were future generations to be allowed for? How could the economic, social and environmental dimensions be considered together? And, as if these were not enough, a cultural dimension was added. All these questions had to be worked out and the looming impact of climate change only made them harder in the new millennium. They were of intense and ongoing interest to economists, some of whom examined the ideal of sustainable forest management.[20]

The forests were not only biologically and socially complex, they were contentious; and the problem of balancing the needs of the present and future in the foresters' principle of sustained yield (Chapter 4) was cited by economists as the precedent for their work. As they investigated the economics needed for sustainable management, they found an increasing number of differences from their neo-classical foundations.[21] These covered not only the difference in the basic assumptions mentioned above, but also issues such as the difference between the quantitative rational models they had constructed and the ecological and process models they would need. Their models had assumed perfect market information, but ecological and climate information would be incomplete and changing. They found many other differences, of which those over discounting and over the objective of models are mentioned here.

The difficulties with discounting were well known to the foresters. They had largely ignored the economic models in practice because the burden of debt from compound interest was more than the natural forests could bear (Chapter 3). The economists came to realise this, as they proposed either not discounting at all, discounting at a negative rate to give greater weight to the future than the present or using some gradually reducing rate.

The greatest difference came with the realisation that the 'optimising' models that they had used to find the most efficient solution were no longer applicable; as there was no single objective, there was no a single 'best' solution to be found. Rather, there were likely to be several solutions and the economists would need the 'satisficing' methods of operations research to find them.

Sustainable forest management as process

It became clear from all this questioning in the new millennium that it was finding the best *process* of seeking a sustainable forest management system that was important, rather than an endpoint that could never been defined. The foresters would never know if they had achieved sustainable forest management, but they could know that they had adopted good processes to do so. What they needed to see the way forward were good examples.[22]

Throughout the world, in different ways, they were trying to develop them in their hopeful, uncertain journey for the new millennium.

Notes

1. Kuhn 1970. p. 19.
2. See for example Ciancio and Nocentini 2000 and Puettmann, Coates and Messier 2008.
3. IUFRO European Congress, Warsaw 2007, *Forests and Forestry in the Context of Rural Development* stated that:
   - Non-wood products, environmental services, and social values are presently not adequately addressed by forest science.
   - Issues such as biodiversity cannot be solved without understanding the social and economic factors that influence human decisions.
   - Two-thirds of issues identified in terms of European forests need socioeconomic research, only 10 per cent of the existing research is presently dedicated to it.
   - Rural development needs to be engaged in the community of place, and to look for new ways to engage these communities as stakeholders in research.
4. Hecht and Cockburn 1990.
5. Indigenous rights and knowledge were also recognised in the UN Convention to Combat Desertification, UNESCO's World Heritage Convention and its MAB Programme, the United Nation Forum on Forests, FAO's programme on Globally Important Indigenous Agricultural Heritage Systems, the Intergovernmental Committee on Intellectual Property and Genetic Resources, Traditional Knowledge and Folklore and in the UN Working Group on Indigenous Populations, the International Covenant on Economic, Social and Cultural Rights.
6. Knudtson and Suzuki published their influential *Wisdom of the Elders* in 1992. It drew on seventeen cases from around the world.
7. Williams 2003, p. 336 and elsewhere, is particularly scathing about Edenic notions of indigenous societies living in harmony with nature.
8. Cited by Trosper and Parrotta 2012.
9. XXII IUFRO World Congress, 8–13 August 2005, Brisbane, Australia, Resolution 2 'Promoting Science for Decision Making' noted that, 'despite the scientific advances so far, the understanding of forest ecosystem dynamics and their relation to continuously changing human demands and global developments such as population growth, migration, urbanization, technology changes and climate change remains incomplete, and that there continues to be a need for advancement of forest-related [traditional] scientific knowledge'. A Task Force on Traditional Forest Knowledge was established by IUFRO (on behalf of the Collaborative Partnership on Forests of which it is a member) in 2005. An international conference on 'Cultural Heritage and Sustainable Forest Management: the Role of Traditional Forest Knowledge' held in 2006 resulted in the research reported in Parrotta and Trosper 2012. In 2010 the Task Force was integrated into the Research Group 9.03 'Forest History and Traditional Knowledge'.

10. Trosper *et al.* 2012.
11. Dowie 2009 describes the 'conservation refugees' evicted from protected areas and some of the reaction against this practice.
12. In 1988 the First International Congress of Ethnobiologists was held in Belém, Brazil. In 1990 it formed its Global Coalition for Biocultural Diversity.
13. An early example is given in a note by the human biologist, Adela Baer in 1989.
14. The lead was taken by Terralingua, an NGO established in 1996.
15. Loh and Harmon 2005.
16. Chazdon *et al.* 2009 provide an overview.
17. Agnoletti *et al.* 2006 set out the guidelines.
18. The 2003 UNESCO *Convention for the Safeguarding of the Intangible Cultural Heritage* covers traditional rural knowledge and practices. The 1995 *Alpine Convention*, established by eight Alpine states (Austria, France, Germany, Italy, Liechtenstein, Monaco, Slovenia and Switzerland) and the EU in 1995, raises awareness of the effects of demographic change on living and working conditions in the Alpine areas and its 2006 Declaration *Population and Culture* covers socio-economic and socio-cultural aspects. Local people are considered to be the main pillar of a self-determining economic development, with an obvious association with tourism. In Austria cultural tourism is one of the prevalent sectors. Initiatives that promote sustainable landscape development include the Netzwerk *Land* (Network *rural areas*), Netzwerk *Forst und Kultur* (Network *forest and culture*) and *Allianz in den Alpen* (Alliance in the Alps, which joins communities and regions of seven states of the Alpine arc).
19. World Commission on Environment and Development, the 'Brundtland Commission' 1987.
20. Pezzey and Thom 2000 and Kant and Berry 2005 provide overviews of the general economic debates on sustainability, the latter focussing on sustainable forest management.
21. Kant 2005 tabulated 26 main differences.
22. A point made by Pezzey and Thom 2000.

CONSIDERING

We set out in writing this book to tell the story of the hopeful science and trusting art of forestry. It is a story about the hopes of foresters and other forest scientists to understand the forests more deeply and it is about their unspoken trust that their knowledge could ensure an enduring sylvan future. Here, in ending our story, we consider how far their hopes were realised and we ask whether their trust was naïve or warranted. How far did their scientific understanding enable a sylvan future? What, over the three centuries discussed in this book, were their successes and failures? And now what might the future hold for forest science and its application?

Ours is only one of the many stories that are and should be told about the forests. Many still lie hidden in the long shelves of textbooks, the burgeoning pile of scholarly papers – over 13,000 with 'forest' in their title were published last year alone – and in the government reports of every country. And they come in different disciplines and political perspectives. We have tried to bridge the gap between the science and the histories, but ours is no less a singular story than others, formed as it is by our lives working in different forests, organisations and languages and carrying our implicit perspective on the careful use of forests and their importance.

At the start of this book we noted some of its limits: we excluded the colourful history of logging technology for example and, in our focus on forest science, we gave scant attention to political economy, beyond setting its context for each section. However, much of what science was done, and what influence it had, was determined by the political and economic forces of time and place – that is a book for others to write – but they governed much of what we consider as the successes and failures of forest science. These have to be considered in the context of their times. Although we have limited the book to the modern scientific era, the change of scale since the seventeenth century has been so great, so overwhelming, for the forests as to make comparisons across the whole period difficult. To reflect on this: there are now one thousand times more people, emitting three thousand times more carbon than in the seventeenth century; empires have come and gone; the world economy, polity, culture, science and technology are increasingly globalised and urbanised. Nevertheless, we have told a story of persistent endeavours to understand the complexity of the forests, to find ways to manage them and to deal with the paradoxes between understanding and managing.

As research became more intricate, the researchers became more specialised. At the start of this book, we discussed how John Evelyn in the seventeenth century had collated traditional English knowledge about growing trees and had made simple trials on his estate; whereas in the twenty-first century it needed groups of different scientists with sophisticated instruments and mathematical models to study the interactions between forests and the climate. The long trend to specialisation brought its rewards in detailed knowledge, but made it difficult to integrate it in the way needed for sustainable forest management, a difficulty made all the harder by the divides between the paradigms for management, biodiversity and social forestry. The place of foresters in this trend changed. From the early nineteenth century to the late twentieth century they had managed the different interests, with wood production as their major concern, personally in daily practice and formally in long-term plans for each forest. Their professional cohesion broke up into the various forms of managing forests, although they kept their managerial role in the multiple-use of the native forests, while being supported or regulated by wildlife, soil, water and other specialists. They took a similar role in the large industrial plantations where they implemented the work of geneticists, tree nutritionists, and soil scientists.

The biodiversity and social forestry forms differed from the multiple-use forests and the industrial plantations in two ways. Firstly, the foresters had no place in the management of the exclusive conservation reserves; those were the domains of ecologists, botanists and wildlife specialists. And they had only a subsidiary place as advisors in the community forests; those were primarily the domains of sociologists, agronomists and human geographers. Secondly, whereas the multiple-use forestry and industrial plantation forms were defined by their vegetation, the other forms had wider scopes: forests were only one of the ecosystems to be conserved for their biodiversity and they supplied only part of the subsistence of local communities.

Forest science in the seventeenth and eighteenth centuries shows some of issues that reappear in different contexts throughout this book. It was clearly directed to the economic and political imperatives of the day: the provision of timber for the navies and the restoration of degraded forests for wood production. It provided the knowledge needed for establishing plantations in Britain and for regenerating forests in France. In Britain it was taken up by large estate owners, but largely ignored by the state, while in

France progress was disrupted for many years by the revolution and its after effects. In both countries knowledge was advanced, but it was governance, not science, that determined the degree of its success.

Forest science was founded in its modern form as a discipline with its own academic standing and research institutes in the early to mid-nineteenth century when industrial capitalism was rapidly expanding, the doctrines of *laissez faire* were taking hold and the old regimes were dead or dying. Notably, it appeared almost as a spin-off from the cameralist drive to put governance on a rational footing. It was an expansive era. Forest science, based on the classification, measurement and experimental principles of other sciences, expanded steadily and started to have its internal specialisations for silviculture, measurement, planning, protection and so forth. It was an era of great achievements, most notably in enunciating the principle of sustained yield and putting forest management on a quantified basis to achieve it. But it was also an era of great simplification, partly by necessity – and partly by choice.

Simplification, as in the search for explanatory models and general laws, is an essential part of any science and can be sought through statistical measures to make the variation of the natural world intelligible. However, the forest scientists' reach into understanding the forests' complexity was limited, even for just the production of wood. Although they developed quite robust statistical models to estimate the quantity and growth of even-aged stands of a single species, mostly conifers, they were much less successful in mixed-age, mixed species forests, especially in the tropics. An exception was in some European forests that they measured intensively and managed by uneven-aged selection systems.

Physical simplification was an essential part of control. It started with the way the foresters mapped the forests, divided them into numbered sections, 'compartments', and identified patches as 'stands' that they considered uniform for practical purposes. Then they imposed orderly arrangements by working systematically through the forests, although they could never make forests as neat as their textbook diagrams.

Biological simplification occurred when the different ages and species in most forests were replaced with orderly patterns of even-aged stands. This was most severe when species were introduced from other places or countries to replace those naturally there. From the end of the nineteenth century, some European foresters reacted against this by calling for a more diverse nature-based forestry that they developed successfully in some regions, and that is being tried elsewhere.

Simplification and the reaction to it provide an example of success, failure and adaptation in forest science. The scientists had advanced their silvicultural knowledge to be able to restore degraded forests and introduce new tree species successfully. However, the resulting simplification was also failure in terms of production in some forests where the soil was harmed or the stands were damaged by disease or wind. More generally, as scientists advanced their knowledge of forest ecology, the single-minded drive to increase wood production became seen as a failure to preserve biodiversity. In consequence, forest scientists adapted their research into developing the multiple-use and close-to-nature forms of forestry.

The contradictions between raising production and protecting biodiversity became most starkly apparent at the end of the twentieth century. At one extreme were the large, intensively-managed and fertilised clonal plantations of introduced species, such as the fast-growing eucalypt plantations in Brazil; at the other extreme were the biodiversity reserves such as those in Tanzania, from which people were evicted. In between were all the public, private and community forests that people were using, managing, or abusing in different ways as they dealt – or failed to deal – with the complementary, competing and conflicting uses and values, and the relative power of their advocates.

Contradictions of scale and scope pervaded modern forest science from its foundation. At the forest level, the foresters had prepared growth and yield models to find the best way to manage each stand and other models to show them how best to regulate a forest to sustain its yield; but the two types were compatible only in the hypothetical ideal of a 'normal' forest. The contradiction could only be overcome by abandoning one or both of the 'bests' and finding a compromise. On a national scale there was a similar contradiction between regulating the yield from each forest separately and providing for regional or national needs. Compromise could be forced by price changes in national markets, by importing or exporting on world markets, or too often by over-riding long-term forest management for short-term expediency: all matters more for economics and politics than forest science.

It is on the world scale that the contradictions of scale became most apparent. In this book we described the development of forestry at the imperial scale and the expansion of forest science to investigate colonial forests and conserve game animals. In the post-colonial era, forest science expanded in many countries, aided by international organisations – mainly FAO, IUFRO, IUCN – and development aid agencies. In the environmental era, forest science started to take on a global scale as it investigated the extent of deforestation and species extinction. Now, in the new millennium, the global scale of climate change has engaged forest scientists in the most demanding investigations that they have ever attempted, but it jams them in the contradiction between what they recognise as needed and what can be achieved politically, either globally or locally.

The contradictions of social scope show another history. Forest science, like most physical and natural sciences, had advanced in the dressage of its aloofness from social and political concerns. However, in applying it, the foresters had regulated the users and policed the boundaries, carrying at times the militaristic and imperial mien of social control. In the environmental era, not only was their knowledge challenged by other disciplines, their authority was sapped by democratic demands for participation and indigenous claims for rights. Such contradictions could not be suppressed in the public forests, but they were studied by social scientists, knowledge was increased and compromises made.

In the new millennium, the global contradictions became bleakly apparent when the United Nations held its 'Earth Summit, Rio+20' conference in 2012 to address the issues of sustainable development.[1] There was no failure of science or of knowledge and all the high rhetoric of the first Earth Summit twenty years before was 'reaffirmed', but no international agreement on either climate or forests could be negotiated; the divide between the economic and political interests of developed and developing countries was too great. Nevertheless, there are other institutions that draw on scientific knowledge as they struggle for limited compromises.

Returning to our opening questions: there is no doubt that the hopes of foresters and other forest scientists to understand the forests more deeply have been rewarded over the centuries, even though their quest can never end. With the benefit of hindsight, we might deem some of their developments to have been misguided, others to have failed and most research to have been superseded by later work, as is its nature. At first sight, we might deem their trust that greater knowledge would secure a future for the forests to have been naïve, but the converse view – that it had no effect – would be equally naïve. More sensibly, forest science and its application can be seen as unevenly developed. For example, European forest scientists have extensive knowledge of their natural forests, and a long history of applying it successfully; and in the industrial plantations, others have extensive knowledge on which they have built their highly productive, intensive silviculture. By contrast, the base of knowledge is generally less developed in the tropics where the forests are more complex and governance is weaker. Moreover, forest science is both enduring and transient. In the long span over which forests have to be considered, the findings of scientific research have accumulated in libraries and archives, even as they are being revised. They may or may not be applied; they may be altered or swept away by political power, or now overtaken by the problems of a changing climate.

It would be comforting to end this book with a celebration of what has been achieved, or the confidence that the future of the world's forests is assured. We cannot; the world is too turbulent and the future too uncertain for that. What we can be sure of is that the hope to understand the complexity of the forests will continue. It will not be enough, but it is not inconsiderable.

Note

1. United Nations Conference on Sustainable Development 2012.

APPENDIX: LIST OF SPECIES

Common name	Scientific name (Syn. = synonym Spp.=species)

Northern hemisphere conifers or softwoods

Silver fir, European	*Abies alba*
Grand fir	*Abies grandis*
Noble fir	*Abies procera*
Lawsons cypress	*Chamaecyparis lawsoniana*
Japanese cedar	*Cryptomeria japonica*
Cunninghamia or Chinese fir	*Cunninghamii lanceolata*
Common juniper	*Juniperus communis*
Larch	*Larix decidua*
Japanese larch	*Larix kaempferi* (syn. *L. leptolepis*)
Hybrid larch	*Larix* x *Eurolepis*
Norway or European spruce	*Picea abies*
Eastern red spruce	*Picea rubens*
Sitka spruce	*Picea sitchensis*
Western white pine	*Pinus albicaulis*
Swiss stone pine or arolla	*Pinus cembra*
Lodgepole pine	*Pinus contorta*
Slash pine	*Pinus elliottii*
Sugar pine	*Pinus lambertiana*
Black pine	*Pinus nigra*
Austrian black pine	*Pinus nigra*, var. *austriaca*
Corsican pine	*Pinus nigra*, var. *maritima*
Pinus patula or Mexican weeping pine	*Pinus patula*
Maritime or cluster pine	*Pinus pinaster*
Ponderosa pine	*Pinus ponderosa*
Radiata or Monterey pine	*Pinus radiata* (syn. *P. insignis*)
Chir pine	*Pinus roxburghii*
Scots pine	*Pinus silvestris*

Eastern white pine or Weymouth pine	*Pinus strobus*
Loblolly pine	*Pinus taeda*
Douglas fir	*Psudeotsuga mensezii* (syn. *Picea douglasii*)
Western red cedar	*Thuja plicata*
Western hemlock	*Tsuga heterophylla*

Southern hemisphere conifers or softwoods

Klinki pine (New Guinea)	*Araucaria hunsteinii*

Northern hemisphere broad leafed trees or hardwoods

Maple	*Acer campestris*
Sycamore	*Acer psuedoplatanus*
Horse chestnut	*Aesculus hippocastanum*
Alder, common or black	*Alnus glutinosa*
Alder, cordate or Italian	*Alnus subcordata*
Birch, pubescent	*Betula pubescens*
Hornbeam	*Carpinus betulus*
Chestnut	*Castanea sativa*
Hazel	*Corylus avellana*
Beech	*Fagus sylvatica*
Ash, European	*Fraxinus excelsior*
Ash, Himalayan	*Fraxinus griffithii*, syn. *F. floribunda*
Walnut, American black	*Juglans nigra and J. Californica*
Walnut	*Juglans regia*
Sweetgum	*Liquidamber styraciflua*
Poplar	*Populus alba*
Aspen	*Populus tremuliodes*
Wild cherry	*Prunus avium*
Oak, Turkey	*Quercus cerris*
Oak, holm	*Quercus ilex*
Oak, sessile	*Quercus petraea*, syn. *Q. sessiliflora*)
Oak, English	*Quercus robur*
Common acacia	*Robinia psuedoacacia*

Willow	*Salix* spp.
Sal	*Shorea robusta*
Rowan (British mountain ash, Quick-beam)	*Sorbus aucuparia*
Whitebeam	*Sorbus aria*
Lime or Linden	*Tilia cordata*
Elm, English	*Ulmus procera* (syn *U. campestris*)

Southern hemisphere and tropical broad leafed trees or hardwoods

Acacia mangium	*Acacia mangium*
Dipterocarps, a large family of 17 genera and 500 species	*Dipterocarpaceae*
River red gum	*Eucalyptus camaldulensis*
Mountain white gum	*Eucalyptus dalrympleana*
Rainbow gum	*Eucalyptus deglupta*
Alpine ash	*Eucalyptus delagatensis*
Blue gum	*Eucalyptus globulus*
Flooded gum	*Eucalyptus grandis*
Mountain ash, Australian	*Eucalyptus regnans*
Swamp mahogany	*Eucalyptus robusta*
Sydney blue gum	*Eucalyptus saligna*
Forest red gum	*Eucalyptus tereticornis*
Timor mountain gum	*Eucalyptus urophylla*
Gmelina	*Gmelina arborea*
Mahogany, African	*Khaya senegalensis*
Ilimo (Papua)	*Octomeles sumatra*
Sal	*Shorea robusta*
Mahogany (West Indies and Honduras)	*Swietania mahagoni*, and *S.macrophylla*
Teak	*Tectona grandis*
Cedar, Australian red	*Toona ciliata*

BIBLIOGRAPHY

(* items freely available on the Internet)

Academy of Sciences [Austrian]. 1989. *Photooxidantien in der Atmosphäre. Luftqualtitäskriterien Ozon.* [Photooxidants in the Atmosphere. Air Quality Criterias Ozone] (Vienna: Bundesministerium für Umwelt Jugend und Familie).

Agnoletti, Mauro, Steven Andersen, Elisabeth Johann, Mart Kulvic, Andrev V. Kushlin, Peter Mayer, Cristina Montiei Molina, John Parrotta, Ian D. Rotherham and Erini Sarasati. 2006. 'The Introduction of Historical and Cultural Values in the Sustainable Management of European Forests'. *Global Environment* 2: 173–99.

Ainsworth, Elisabeth A. and Stephen P. Long. 2005. 'So what have we learned from 15 years of free-air CO_2 enrichment (FACE)? A meta-analytic review of the responses of photosynthesis, canopy properties and plant production to rising CO_2'. *New Phytologist* **165**: 351–72.

Albion, R.G. 1926. *Forests and Sea Power: the Timber Problem of the Royal Navy 1652–1862.* (Cambridge, Mass: Harvard University Press).

Anderson, Mark L. 1967. *A History of Scottish Forestry.* (London: Nelson).

Anderson, Mark L. 1932. *The Natural Woodlands of Britain and Ireland.* (Oxford: Holywell Press).

Anderson, Mark L. 1950. *The Selection of Tree Species: an Ecological Classification for Conditions Found in Great Britain and Ireland.* (Edinburgh: Oliver and Boyd).

Anon. 1813. 'Sylva, or a Discourse of Forest trees ... with notes by H. Hunter, fourth edition, 1812', *Quarterly Review* 9(17): 45–57.

Anon. 1832. 'Account of the Larch Plantations on the Estates of Atholl and Dunkeld, Executed by the Late John, Duke of Atholl'. *Transactions of the Highland and Agricultural Society of Scotland*: 165–219.

Armin. n.d. *Provenienzforschung – eine wichtige Aufgabe für die Forstwirtschaft* [Provenance Research – an Important Task for Forestry]. (Großhansdorf & Waldsieversdorf, Germany: Institut für Forstgenetik und Forstpflanzenzüchtung [Institute of Forest Genetics and Forest Tree Breeding).*

Arnold, J.E.M. 1998. *Managing Forests as Common Property.* (Rome: FAO).*

Arnold, J.E.M. 2001. *Forests and People: 25 Years of Community Forestry.* (Rome: FAO).*

Arnold, J.E.M. 2004. 'Common Property Resource Management', in Jeffery Burley, Julian Evans and John A. Youngquist (eds.) *Encyclopedia of Forest Sciences.* (Amsterdam: Elsevier), pp. 1131–6.

Baer, Adela. 1989. 'Maintaining Biocultural Diversity'. *Conservation Biology* 3(1): 97–98.

Baker, F.S. 1934. *Theory and Practice of Silviculture.* (New York: McGraw).

Baker, F.S. 1950. *Principles of Silviculture.* (New York: McGraw).

Baker, Susan C. and Steve M. Read. 2011. 'Variable Retention Silviculture in Tasmania's Wet Forests: Ecological Rationale, Adaptive Management and Biodiversity Benefits'. *Australian Forestry* **74**(3): 218–32.

Barber, Klaus H. and Susan A. Rodman. 1990. 'FORPLAN: the Marvellous Toy'. *Journal of Forestry* **88**(5): 26 –30.

Barrère, Pierre, Robert Heisch and Serge Lerat. 1962. *La Règion du Sud–Ouest* (Paris: Presses Universitaires de France), pt 2, ch. 5, 'La Forêt Landaise'.

Barton, Gregory Allen. 2002. *Empire Forestry and the Origins of Environmentalism.* (Cambridge University Press).

Baule, Hubert and Claude Fricker (trans. C.L. Whittles). 1970. *The Fertilizer Treatment of Forest Trees.* (Munich: BLV Verlagesellschaft).

Bechmann, Roland (trans. Kathryn Durham). 1990. *Trees and Man: the Forest in the Middle Ages.* (New York: Paragon House), first published 1984 as *Des Arbres et des Homes: La Forêt au Moyen Age.* (Paris: Flammarion).

Bendix, Bernd. 2008. *Geschichte der Forstpflanzenzucht in Deutschland von ihren Anfängen bis zum Ausgang des 19. Jahrhunderts* [History of Forest Tree Breeding in Germany from its beginnings to the end of the 19th Century]. (Remagen-Oberwinter: Kessel).

Binkley, Don. 1986. *Forest Nutrition Management.* (New York: Wiley).

Bitterlich, Walter. 1984. *The Relascope Idea: Relative Measurements in Forestry.* (Farnham Royal: Commonwealth Agricultural Bureau).

Blake, William. 1794. 'The Tyger', from *Songs of Innocence and Experience.**

Böhmerle, Karl. 1900. *Bisherige Erfahrungen aus einigen Durchforstungs- und Lichtungsversuchsflächen der Forstlichen Versuchsanstalt Mariabrunn; anlässlich der Pariser Weltausstellung 1900* [Results from Some Thinning Plots of the Imperial Forest Research Institute in Mariabrunn: Presented at the Paris World Exhibition in 1900]. (Vienna: Frick).

Bolin, Bert. 2007. *A History of the Science and Politics of Climate Change: The Role of the Intergovernmental Panel on Climate Change.* (Cambridge University Press).

Bonnaire, P. 2000. 'Il y a Trois Cent Ans Naissait Duhamel du Monceau (1700–1782) [Three Hundred Years Since the Birth of Duhamel du Monceau (1700–1782)]. *Revue Forestière Français* 2000 (2): 145–58.*

Boomgaard, Peter. 1988. 'Forests and Forestry in Colonial Java, 1677–1942', in John Dargavel, Kay Dixon and Noel Semple (eds.) *Changing Tropical Forests: Historical Perspectives on Today's Challenges in Asia, Australasia and Oceania.* (Canberra: Center for Resource and Environmental Studies, The Australian National University).

Bourke, Max. 2011. 'Private Investment in Biodiversity Conservation: a Growing Trend in the Western World?' *Global Environment.* **7–8**: 39–49.

BRACELPA, Associaçao Brasileira de Cellulose e Papel [Brazilian Association of Pulp and Paper Manufacturers]. 2011. *Brazilian Pulp and Paper Industry.* (São Paulo: BRACELPA).*

Bradsock, Ross A., A. Malcolm Gill and Richard J. Williams. 2012. *Flammable Australia: Fire Regimes, Biodiversity and Ecosystems in a Changing World.* (Collingwood, Vic: CSIRO Publishing).

Bibliography

Brandis, Dietrich. 1906. *Indian Trees: an Account of Trees, Shrubs, Woody Climbers, Bamboos and Palms Indigenous or Commonly Cultivated in the British Indian Empire.* (London: Constable).*

Brandl, H. 1992. 'Entwicklungslinien in Deutscher Forstwirtschaft und Deutscher Forstwissenschaft mit Internationaler Ausstrahlung' [Trends in German Forestry and Forest Science with International Implications], in *IUFRO Proceedings Centennial [Conference] 31 Aug–4 Sep 1992* (Vienna: IUFRO), pp. 43–72.

Brown, Chris, Patrick B. Durst and Thomas Enters. 2005. 'Perceptions of Excellence: Ingredients of Good Forest Management', in Patrick B. Durst, Chris Brown, Henrylito D. Tacio and Miyuki Ishikawa (eds.) *In Search of Excellence: Exemplary Forest Management in Asia and the Pacific.* (Bangkok: FAO Regional Office for Asia and the Pacific).*

Brown, John Croumbie. 1883. *French Forest Ordinance of 1669; with Historical Sketch of Previous Treatment of Forests in France.* (Edinburgh: Oliver and Boyd).*

Brown, John Croumbie. 1887. *Schools of Forestry in Germany, with Addenda Relative to a Desiderated British National School of Forestry.* (Edinburgh: Oliver and Boyd).*

Bruce, D. and F. Schumacher. 1930, 1935. *Forest Mensuration.* (New York: McGraw-Hill).

Buchy, Marlène. 1996. *Teak and Arecanut: Colonial State, Forest and People in the Western Ghats (south India) 1800–1947.* (Pondichéry: Institut Français de Pondichéry – Indira Ghandi National Centre for the Arts).

Burdon, Rowland D. and William J. Libby. 2006. *Genetically Modified Forests: From Stone Age to Modern Biotechnology.* (Durham, NC: Forest History Society).

Burley, J. 2004. 'A Historical Overview of Forest Tree Improvement', in Jeffery Burley, Julian Evans and John A. Youngquist (eds.) *Encyclopedia of Forest Sciences.* (Amsterdam: Elsevier), pp. 1532–8.

Burton, John A. 1987. 'Bibliography of Red Data Books (Part 1, Animal Species)'. *Oryx* **18**: 61–4.

Campbell-Culver, Maggie. 2001. *The Origin of Plants: the People and Plants that have Shaped Britain's Garden History since the Year 1000.* (London: Headline Book Publishing).

Campbell-Culver, Maggie. 2006. *A Passion for Trees: the Legacy of John Evelyn.* (London: Transworld Publishers).

Cardigan, Earl. 1949. *The Wardens of Savernake Forest.* (London: Routledge & Kegan Paul).

Carron, L.T. 1985. *A History of Forestry in Australia.* (Canberra: Australian National University Press).

Carruthers, Jane. 1997. 'Nationhood and National Parks: Comparative Examples from the Post-Imperial Experience', in Tom Griffiths and Libby Robin (eds.) *Ecology and Empire: Environmental History of Settler Societies* (Melbourne University Press).

Carter, Neil H., Binoj K. Shrestha, Jhamak B. Karki, Narendra Man Babu Pradhan and Jianguo Liu. 2012. 'Coexistence between wildlife and humans at fine spatial scales'. *PNAS: Proceedings of the National Academy of Sciences of the United States of America* **109**(38): 15360-5.*

Chakrabarti, Milindo, Samar K. Datta, E. Lance Howe and Jeffrey B. Nugent. 2005. 'Joint Forest Management: Experience and Modelling', in Sashi Kant and R. Albert Barry (eds.) *Economics, Sustainability, and Natural Resources: Vol. 1 Economics of Sustainable Forest Management.* (Dordrecht: Springer).

Chakrabati, Ranjan. 2009. 'Local People and the Global Tiger: An Environmental History of the Sundarbans'. *Global Environment* **3**: 72 –95.

Chambrelent [Jules]. 1887. *Les Landes de Gascogne.* (Paris: Libraire Polytechnique, Baudry et Cie).*

Chapman, H.H. 1921. *Forest Mensuration.* (New York: Wiley).

Chapman, H.H. and D.B. Demeritt. 1931, 1936. *Elements of Forest Mensuration.* (Albany, N.Y: Williams Press).

Chapman, H.H. and W.H. Meyer. 1949. *Forest Mensuration.* (New York: McGraw-Hill).

Chappell, H.N., D.W. Cole, S.P. Gessel and R.B. Walker. 1991. 'Forest fertilization research and practice in the Pacific Northwest'. *Fertilizer Research* [re-titled *Nutrient Cycling in Agroecosystems*] **27**: 129–40.

Chazdon, Robin L., Celia A. Harvey, Oliver Komar, Daniel M. Griffith, Bruce G. Ferguson, Miguel Martínez-Ramos, Helda Morales, Ronald Nigh, Lorena Soto-Pinto, Michiel van Breugel and Stacy M. Philpott. 2009. 'Beyond Reserves: A Research Agenda for Conserving Bodiversity in Human-Modified Tropical Landscapes'. *Biotropica* **41**(2): 142–53.

Chundawat, R.S., B. Habib, U. Karanth, K. Kawanishi, J. Ahmad Khan, T. Lynam, D. Miquelle, P. Nyhus, S. Sunarto, R. Tilson and Sonam Wang. 2011. *Panthera tigris*, in IUCN 2011. IUCN Red List of Threatened Species. Version 2011.2.*

Ciancio, O. and S. Nocentini. 2000. 'Forest Management from Positivism to the Culture of Complexity', in M. Agnoletti and S. Anderson (eds.) *Methods and Approaches in Forest History.* (Wallingford: CABI Publishing).

Clutter, J.L. 1968. *MAX-MILLION–a Computerized Forest Management Planning System.* (Athens, Georgia: School of Forest Resources, University of Georgia).

Clutter, Jerome L., James C. Fortson, Leon V. Piennar, Graham H. Brister and Robert L. Bailey. 1983. *Timber Management: a Quantitative Approach.* (New York: John Wiley).

Conwentz, Hugh. 1904. *Die Gefährdung der Naturdenkmäler und Vorschläge zu ihrer Erhaltung* [The Endangerment of the Natural Monuments and Suggestions for their Conservation]. (Berlin: Gebrüder Bornträger).

Cronon, William. 1996. 'The Trouble with Wilderness or Getting Back to the Wrong Nature', in William Cronon (ed.) *Uncommon Ground: Rethinking the Human Place in Nature.* (New York: W.W. Norton).

Crosby, Alfred W. 1986. *Ecological Imperialism: the Biological Expansion of Europe, 900–1900.* (Cambridge University Press).

Dargavel, J., J. Holden, R.P. Brinkman and B.J. Turner. 1995. 'Advanced Forest Planning: Reflections on the Otway Forest Management Planning Project', *Australian Forestry* **58**(2): 21–30.

Dargavel, J.B. 1976. 'Evaluating the Role of Thinning in Development Forestry'. *New Zealand Journal of Forestry Science* **41**(8): 242–52.

Dargavel, John. 2006. 'From Exploration to Science: Lane Poole's Forest Surveys of Papua and New Guinea, 1922– 1924'. *Historical Records of Australian Science* **17**: 71– 90.

Dargavel, John. 2008. *The Zealous Conservator: A Life of Charles Lane Poole.* (Crawley: University of Western Australia Press).

Dargavel, John. 2010. 'Netting the Global Forests: Attempts at Influence'. *Global Environment* **5**: 126–57.

Darley, Gillian. 2006. *John Evelyn, Living for Ingenuity.* (New Haven, Conn: Yale University Press).

Darrow, W. 1977. *David Ernest Hutchins: A Pioneer in South African Forestry.* (Pretoria: Department of Forestry).

Dawkins, H.C. and M. S. Philip. 1998. *Tropical Moist Forest Silviculture and Management: a History of Success and Failure.* (Wallingford: CAB International).

Dieterich, V. 1944. 'Vorbedingungen bestmöglichen Ausgleichs zwischen angespanntestem Holzbedarf und nachhaltiger Forstbewirtschaftung' [Preconditions for the Best Balance between intense demand for wood and sustainable forestry]. *Mitteilungen der Hermann-Göring-Akademie der Deutschen Forstwirtschaft* 4. Jahrgang volume I 1944; special issue (Frankfurt am Main: J.D. Sauerländer).

Dimitz, L. 1907. 'Gesetzliche Vorkehrungen betreffend den Schutz der natürlichen Landschaft und die Erhaltung der Naturdenkmäler. Die Erhaltung ursprünglicher Waldbestände; Vorschläge zur freiwilligen, administrativen und legislativen Mitwirkung' [Statutory Provisions Relating to the Protection of the Natural Monuments; the Preservation of Original Forests; Proposals for Voluntary, Administrative and Legislative Participation]. 8. Intern. Land- und forstw. Kongress [8th International, Agricultural and Forestry Congress], Wien 21–25. Mai 1907. *Referate* [Speeches] VIII–IX, Bd. 4.

Doughty, Robin W. 2000. *The Eucalyptus: a Natural and Commercial History of the Gum Tree.* (Baltimore: John Hopkins University Press).

Douglas, David. 1914. *Journal kept by David Douglas During his travels in North America 1823–1827 : together with a Particular Description of Thirty-Three Species of American Oaks and Eighteen Species of Pinus, with Appendices Containing a List of the Plants introduced by Douglas and an Account of His Death in 1834.* (London: W. Wesley & Son under the Direction of the Royal Horticultural Society).*

Dowie, Mark. 2009. *Conservation Refugees: the Hundred-year Conflict Between Global Conservation and Native Peoples.* (Cambridge, MA: Massachusetts Institute of Technology).

Duhamel du Monceau, M. [Henri-Louis]. 1969 [1760]. *Semis et Plantations des Arbres et de Leur Culture.* (New York: Readex Microprint [Paris: Guerin & Delatour]).*

Ebermayer, G. 1891. 'Die Waldstreufrage. Vortrag gehalten am 23. Jan. 1891 in der bayer. Gartenbau-Gesellschaft zu München' [The Litter Issue. Lecture Held in the Bavarian Horticultural Society, 23 January 1891], in *Forstlich-naturwissenschaftliche Zeitschrift*, 2.

Edwards, A.W.F. 2001. 'Darwin and Mendel United: the Contributions of Fisher, Haldane and Wright up to 1932', in E.C.R. Reeve (ed.) *Encyclopedia of Genetics.* (London: Fitzroy Dearborn), pp. 688–91.

Edwards, M.V. 1956. 'The Hybrid Larch. Larix x europlepis Henry'. *Forestry* **29**(1): 29–43.

Eldridge, Ken, John Davidson, Christopher Harwood and Gerrit van Wyk. 1993. *Eucalyptus Domestication and Breeding.* (Oxford: Clarendon Press).

European Communities. 2003. *Sustainable Forestry and the European Union: Initiatives of the European Commission.* (Luxembourg: Office for Official Publications of the European Communities).

Evelyn, John. [1664] 1995. 'Sylva: or a Discourse of Forest trees, and the Propagation of Timber in his Majesties Dominions, &c. (1st edn.)', in Guy de la Bédoyère (ed.) *The Writings of John Evelyn.* (Woodbridge: Boydell Press).

Evelyn, John. [1706] ed. John Nisbet. 1908. *Silva: or a Discourse of Forest trees, and the Propagation of Timber in his Majesties Dominions, &c. Reprint of 4th edn.* (New York: Doubleday).*

Evers, Fritz-Helmut. 1991. 'Forest Fertilization – Present State and History with Special Reference to South German Conditions'. *Fertilizer Research* [re-titled *Nutrient Cycling in Agroecosystems*] **27**: 71–80.

FAO. 1978. *Forestry for Local Community Development.* (Rome: FAO, Forestry Paper 7).*

FAO. 2010. *Global Forest Resources Assessment 2010.* (Rome: FAO).*

Farjon, A., C.N. Page and the IUCN/SSC Conifer Specialist Group. 1999. *Conifers: Status Survey and Conservation Action Plan.* (Gland, Switzerland: IUCN).*

Faustmann, M. [1849] (trans. W. Linnard). 1995. 'Calculation of the Value which Forest Land and Immature Stands Possess for Forestry'. *Journal of Forest Economics* **1**: 7–44.

Federal Forest Research Centre Vienna (ed.) 1974. *The History of the Federal Forest Research Centre and its Institutes.* (Vienna: Österreichischer Agrarverlag. Mitteilungen der forstlichen Bundes-Versuchsanstalt Wien [Releases of the Federal Forest Research Centre Vienna number] 106).

Fellows, Otis E. and Stephen F. Milliken. 1972. *Buffon.* (New York: Twayne Publications).

Ferriera, J.M. de Aguiar and J.L. Stape. 2009. 'Productivity Gains by Fertilisation in *Eucalyptus urophylla*: Clonal Plantations Across Gradients in Site and Stand Conditions'. *Southern Forests* **71**(4): 253–8.

Fischlin, A., G.F. Midgley, J.T. Price, R. Leemans, B. Gopal, C. Turley, M.D.A. Rounsevell, O.P. Dube, J. Tarazona and A.A. Velichko. 2007. 'Ecosystems, their properties, goods, and services', in M.L. Parry, O.F. Canziani, J.P. Palutikof, P.J. van der Linden and C.E. Hanson (eds.) *Climate Change 2007: Impacts, Adaptation and Vulnerability: Contribution of Working Group II to the Fourth Assessment Report of the Intergovernmental Panel on Climate Change.* (Cambridge University Press), pp. 211–72.*

Fisher, R.A. 1925. *Statistical Methods for Research Workers.* (Edinburgh: Oliver and Boyd).

Fisher, R.A. 1935. *The Design of Experiments.* (Edinburgh: Oliver and Boyd).

Fisher, W.R. 1896. *Schlich's Manual of Forestry, Vol V: Forest Utilization.* [translation and revision of Gayer 1868] (London: Bradbury Agnew).

Foster, John Bellamy. 2000. *Marx's Ecology: Materialism and Nature.* (New York: Monthly Review Press).

Franklin, Jerry F., Dean Rae Berg, Dale A. Thornburgh and John C. Tappeiner. 1997. 'Alternative Silvicultural Approaches to Timber Harvesting: Variable Retention Harvest Systems', in Kathryn A. Kohm and Jerry F. Franklin (eds.) *Creating a Forestry for the 21st Century: the Science of Ecosystem Management.* (Washington: Island Press).

Gammage, Bill. 2011. *The Biggest Estate on Earth: How Aborigines Made Australia.* (Crows Nest, NSW: Allen & Unwin).

Gaughwin, Denise. 1999. 'Species Trials and Arboreta in Tasmania', in John Dargavel and Brenda Libbis (eds.) *Australia's Ever-changing Forests IV: Proceedings of the Fourth National Conference on Australia's Forest History.* (Canberra: Centre for Resource and Environmental Studies, The Australian National University), pp. 329–41.*

Gayer, Karl. 1868. *Die Forstbenutzung* [Forest Utilisation] (Aschaffenburg: n.k.).* [See also translation in Fisher 1896].

Gayer, Karl. 1880. *Der Waldbau* [Silviculture]. (Berlin: Wiengandt).

Gayer, Karl. 1884, 1886. *Die neue Wirthschaftrichtung in den Staatswaldungen des Spessarts* [New Economic Goals in the State Forests of Spessart]. (München: n.k.).

Gayer, Karl. 1886. *Der gemischte Wald, seine Begründung und Pflege, insbesondere die Horst- und Gruppenwirtschaft* [The mixed forest, its setting up and tending]. (Berlin: P. Parey).

Gessel, S.P. 1968. 'Progress and Needs in Tree Nutrition Research in the Northwest', in *Forest Fertilization, Theory and Practice.* (Muscle Shoals, Alabama: Tennessee Valley Authority), pp. 216–25.

Gillbank, Linden. 1993, 'Nineteenth Century Perceptions of Victorian forests: Ideas and Concerns of Ferdinand Mueller', in John Dargavel and Sue Feary (eds.) *Australia's Ever-changing Forests II, Proceedings of the Second National Conference on Australian Forest History.* (Canberra: Centre for Resource and Environmental Studies, Australian National University).*

Gilmour, D.A. and E.J. Fisher. 1991. *Villagers, Forests and Foresters: the Philosophy, Process and Practice of Community Forestry in Nepal.* (Kathmandu: Sahayogi Press).

Grace, John and Mark Rayment. 2000. 'Respiration in the Balance'. *Nature* **404**: 819–20.

Graves, H.S. 1906. *Forest Mensuration.* (New York: Wiley).

Gregory, G. Robinson. 1955. 'An Economic Approach to Multiple Use'. *Forest Science* **1**(1): 6–13.

Grove, Richard H. 1995. *Green Imperialism: Colonial Expansion, Topical island Edens and the Origins of Environmentalism, 1600– 1860.* (Cambridge University Press).

Grut, Mikael. 1965. *Forestry and Forest Industry in South Africa* (Cape Town: A.A. Balkema).

Guha, Ramachandra. 1989. *The Unquiet Woods: Ecological Change and Peasant Resistance in the Indian Himalaya.* (Delhi: Oxford University Press).

Hagner, Stig. 1966. 'Timber Production by Forest Fertilisation'. *Proceedings of the Fertiliser Society* [London]. No. 94: 38–54.

Haig, Irvine T., M.A. Huberman and U. Aung Din. 1957–1958. *Tropical Silviculture.* (Rome: FAO, 3 vols.).

Hartely, Beryl. 2010. 'Exploring and Communicating Knowledge of Trees in the Early Royal Society'. *Notes and Records of the Royal Society* **64**(3): 229–50.*

Hecht, Susanna and Alexander Cockburn. 1990. *The Fate of the Forest: Developers, Destroyers and Defenders of the Amazon.* (London: Penguin).

Hendry, George R., David S. Ellsworth, Keith F. Lewin and John Nagy. 1999. 'A Free-air Enrichment System for Exposing Tall Forest to Elevated Atmospheric CO_2'. *Global Change Biology* 5: 293–309.*

Heyder, J. Chr. 1983. *Waldbau im Wandel. Zur Geschichte des Waldbaus von 1870 bis 1950 dargestellt unter besonderer Berücksichtigung der Bestandesbegründung und der forstlichen Verhältnisse Norddeutschlands* [Forestry in transition. Presented on the history of silviculture from 1870 to 1950 with particular emphasis on stand establishment and forest conditions in northern Germany]. (Frankfurt am Main: J.D Sauerländers Verlag).

Hiley, W.E. 1930. *The Economics of Forestry.* (Oxford University Press).

Hiley, W.E. 1956. *The Economics of Plantations.* (London: Faber and Faber).

Hobley, Mary. 1996. *Participatory Forestry: The Process of Change in India and Nepal.* (London: Overseas Development Institute).

House, Syd and Christopher Dingwell. 2004. '"A Nation of Planters": Introducing the New Trees, 1650–1900', in T.C. Smout (ed.) *People and Woods in Scotland: a History.* (Edinburgh University Press) pp. 128–57.

Hughes, J. Donald. 1996. *Pan's Travail: Environmental Problems of the Ancient Greeks and Romans.* (Baltimore: John Hopkins University Press).

Hughes, J. Donald. 1997. 'Ancient Forests: the Idea of Forest Age in the Greek and Latin Classics', in John Dargavel (ed.) *Australia's Ever-changing Forests III: Proceedings of the Third National Conference on Australian Forest History.* (Canberra: Centre for Resource and Envrironmental Studies, The Australian National University).*

Humphreys, David. 1996. *Forest Politics: the Evolution of International Cooperation.* (London: Earthscan).

Humphreys, David. 2006. *Logjam: Deforestation and the Crisis of Global Governance.* (London: Earthscan).

Hundeshagen, J.C. 1819. *Methodologie und Grundriß der Forstwirtschaft* [Methodology and Main Features of Forestry]. (Tübingen: H. Laupp).

Hundeshagen, J.C. 1821–1831. *Encyclopädie der Forstwissenschaft, systematisch abgefasst* [Encyclopaedia of Forest Science]. 3 vols (Tübingen: H. Laupp).

Hundeshagen, J.C. 1826. *Die Forstabschätzung auf neuen wissenschaftlichen Grundlagen* [Forest Assessment Based on New Scientific Principles]. (Tübingen: H. Laupp).

Hundeshagen, J.C. 1827–1830. *Lehrbuch der forst- und landwirthschaftlichen Naturkunde* [Textbook of Forestry and Agriculture Natural History]. (Tübingen: H. Laupp).

Hunter, Michael. 1989. *Establishing the New Science: the Experience of the Early Royal Society.* (Woodbridge: Boydell Press).

Husch, B., T.W. Beers, and J.A. Kershaw. 2003. *Forest Mensuration.* (Hoboken, N.Y: Wiley).

Hutchins, D.E. 1905. 'Forestry in South Africa', in W. Flint and J.D.F. Gilchrist (eds.) *Science in South Africa: a Handbook and Review.* (Cape Town: Mathew Miller) pp. 391–413.

Ilvessalo, Lauri. 1927. 'Forest Research Work in Finland', *Forestry* 1: 73–77.

Innes, J.L. 2004. 'Carbon Cycle', in Jeffery Burley, Julian Evans and John A. Youngquist (eds.) *Encyclopedia of Forest Sciences.* (Amsterdam: Elsevier), v.1, pp. 139–44.

Innes, John L. 1993. *Forest Health: its Assessment and Status.* (Wallingford: CAB International).

International Union of Forest Research Organisations. 1992. *100 Years of IUFRO, 1892–1992.* (Vienna: IUFRO).

IUCN. 1997. *Proceedings* [First] *World Conservation Congress.* (Gland, Switzerland: IUCN).*

IUCN and UNEP. 2009. *The World Database on Protected Areas (WDPA).* (Cambridge, UK: UNEP-WCMC).

Jackson, G.E. and J. Grace. 2004. 'Impacts of Elevated CO_2 and Climate Change', in Jeffery Burley, Julian Evans and John A. Youngquist (eds.) *Encyclopedia of Forest Sciences.* (Amsterdam: Elsevier), v.1, pp. 144–52.

James, N.D.G. 1981. *A History of English Forestry.* (Oxford: Basil Blackwell).

Jeffers, J.N.R. 1961. 'The Electronic Digital Computer in Forestry'. *Unasylva* **15**(4).*

Jepson, Paul and Robert J. Whittaker. 2002. 'Histories of Protected Areas: Internationalisation of Conservation Values and their Adoption in the Netherlands Indies (Indonesia)'. *Environment and History* **8**(2): 129–72.

Jerram, M.R.K. 1939, 1949. *Elementary Forest Mensuration.* (London: Thomas Murby).

Johann, E. 2001. 'Zur Geschichte des Natur– und Landschaftsschutzes in Österreich. Historische "Ödflächen" und ihre Wiederbewaldung' [On the History of Nature and Landscape Protection in Austria: Historical "Wasteland" and Reforestation], in Norbert Weigl (ed.) *Faszination Forstgeschichte* [Fascination of Forest History]. (Wien: Schriftenreihe des Instituts für Sozioökonomik der Forst– und Holzwirtschaft) vol. 42, pp. 41–60.

Johann, Elisabeth. 2006. 'Historical Development of Nature-based Forestry in Central Europe', in Juri Diaci (ed.) *Nature-based Forestry in Central Europe: Alternatives to Industrial Forestry and Strict Preservation.* (Ljubljana: Biotechnicka Fakulteta. Studia Slovenica no. 126).*

Johann, Elisabeth, Mauro Agnoletti, Anna-Lena Axelsson, Matthias Bürgi, Lars Östlund, Xavier Rochel, Uwe Ernst Schmidt, Anton Schuler, Jens-Peter Skovsgaard and Verena Winiwarter. 2004. 'History of Secondary Spruce Forests in Europe', in Hubert Sterba, Konstantin Teuffel, Jens Peter Skovsgaard, Emil Klimo, Jörg Hansen and Heinrich Spiecker (eds.) *Norway Spruce Conversion: Options and Consequences.* (Leiden: Brill).

Johann, Elisabeth, Mauro Agnoletti, János Bölöni, Seçil Yurdakul Erol, Kate Holl, Jürgen Kusmin, Jesús García Latorre, Juan García Latorre, Zsolt Molnár and Xavier Rochel, Ian D. Rotherham, Erini Sarasi, Mike Smith, Lembitu Terang, Mark van Bentham and Jim van Laar. 2012. 'Europe', in J.A. Parrotta and R.L. Trosper (eds.) *Traditional Forest-Related Knowledge: Sustaining Communities and Biological Diversity.* (Dordrecht: Springer).

Johannson, Per-Olav and Karl-Gustaf Löfgren. 1985. *The Economics of Forestry and Natural Resources.* (Oxford: Basil Blackwell).

Kanowski, P.J. and N.M.G. Borralho. 2004. 'Economic Returns from Tree Breeding', in Jeffery Burley, Julian Evans and John A. Youngquist (eds.) *Encyclopedia of Forest Sciences.* (Amsterdam: Elsevier) pp. 1561–8.

Kanowski, Peter 2001. 'Trees Still Challenging Breeders at the Millennium', in E.C.R. Reeve (ed.) *Encyclopedia of Genetics.* (London: Fitzroy Dearborn) pp. 688–91.

Kant, Sashi. 2005. 'Post-Newtonian Economics and Sustainable Forest Management', in Kant, Sashi and R. Albert Berry (eds.) *Economics, Sustainability, and Natural Resource Management: Economics of Sustainable Forest Management.* (Dordrecht: Springer).

Kant, Sashi and R. Albert Berry. 2005. 'Economics Sustainability and Forest Management', in Kant, Sashi and R. Albert Berry (eds.) *Economics, Sustainability and Natural Resources: Economics of Sustainable Forest Management.* (Dordrecht: Springer).

Kasthofer, K. 1828. *Der Lehrer im Walde* [The Teacher in the Forest]. 2 parts (Jenni: Berne).

Keeves, A. 1966. 'Some Evidence of Loss of Productivity with Successive Rotations of Pinus Radiata in the South-East of South Australia'. *Australian Forestry* **30**: 52–63.

Keith, H., E. van Gorsel, K.L. Jacobsen and H.A. Cleugh 2012. 'Dynamics of Carbon Exchange in a Eucalyptus Forest in Response to Interacting Disturbance Factors'. *Agricultural and Forest Meteorology* **153**: 67–81.

Keith, H., R. Leuning, K.L. Jacobsen, H.A. Cleugh, E. van Gorsel, R.J. Raison, B.E. Medlyn, A. Winters and C. Keitel. 2009. 'Multiple Measurements Constrain Estimates of Net Carbon Exchange by a *Eucalyptus* Forest. *Agricultural and Forest Meteorology* **149**: 535–58.

Kengen, Sebastião. 1997. *Forest Valuation for Decision-Making: Lessons from Experience and Proposals for Improvement*. (Rome: Food and Agriculture Organization).*

Kent, Brian, B. Bruce Bare, Richard C. Field and Gordon A. Bradley. 1991. 'Natural Resource Land Management Planning Using Large-scale Linnear Programs: the USDA Forest Service Experience with FORPLAN'. *Operations Research* **39**(1): 13–27.

Kimmins, J.P. 1997. *Forest Ecology: A Foundation for Sustainable Management*, 2nd edn. (Upper Saddle River, New Jersey: Prentice-Hall).

Knuchel, H. (trans. M.L. Anderson) [1950 *Planung und Kontrolle im Forstbetreib*]. 1953. *Planning and Control in the Managed Forest*. (Edinburgh: Oliver and Boyd).

Knudtson, Peter and David Suzuki. 1992. *Wisdom of the Elders*. (North Sydney: Allen & Unwin).

Kohn, Marek. 2004. *A Reason for Everything: Natural Selection and the English Imagination*. (London: Faber).

Köstler, Josef (trans. M.L. Anderson). 1956. *Silviculture*. (Edinburgh: Oliver and Boyd).

Kuhn, Thomas S. [1962] 1970. *The Structure of Scientific Revolutions*. (University of Chicago Press).

Kupper, Patrick. 2009. 'Science and National Parks: a Transatlantic Perspective'. *Environmental History* **14**(1): 58–81.

Lamlom, S.H. and R.A. Savidge. 2003. 'A Reassessment of Carbon Content in Wood: Variation Within and Between 41 North American Species'. *Biomass and Energy* **25**: 381–8.

Langlet, Olaf. 1971. 'Two Hundred Years of Genecology'. *Taxon* **20**(5/6): 653–721.

Lawrence, A. and S. Gillet. 2004. 'Joint and Collaborative Forest Management', in Jeffery Burley, Julian Evans and John A. Youngquist (eds.) *Encyclopedia of Forest Sciences*. (Amsterdam: Elsevier) pp. 1143–57.

Leibundgut, H. 1949. 'Grundzüge der schweizerischen Waldbaulehre' [Basics in the Doctrine of Silviculture]. *Forstwissenschaftliches Centralblatt* **68**: 257–291.

Leibundgut, H. 1989. *Naturnahe Waldwirtschaft* [Silviculture close to Nature]. (Siegen, Germany: Wilhelm-Münker-Stiftung).

Lessard, L.G. 1986. 'Importance of Social Forestry in the African Environment: an IDRC View', *WCED Archive Collection*, v.27, doc. 170.*

Leuning, Ray, Helen A. Cleugh, Steven J. Zegelin and Dale Hughes. 2005. 'Carbon and Water Fluxes over a Temperate Eucalyptus Forest and a Tropical Wet/Dry Savanna in Australia: Measurements and Comparison with MODIS Remote Sensing Estimates'. *Agricultural and Forest Meteorology* **129**: 151–73.

Lindenmayer, David B. and Jerry F. Franklin 2002. *Conserving Forest Biodiversity: a Comprehensive Multiscaled Approach.* (Washington: Island Press).

Liocourt, F. de [1898] (trans. Maria Nygren). 2001. 'De l'amanagement des sapinières: On the amelioration of fir forests' [*Bulletin Trimestriel, Société de Franche-Comté de Belfort* 1898: 396–409] (Columbia: University of Missouri-Columbia).*

Loh, Jonathon and David Harmon. 2005. 'A Global Index of Biocultural Diversity'. *Ecological Indicators* **5**(3): 231-41.

Loudon, J.C. 1838. *Arboretum et Fructetum Britannicum; or the Trees and Shrubs of Britain.* (London: Longman).*

Lowood, Henry E. 1990. 'The Calculating Forester: Quantification, Cameral Science, and the Emergence of Scientific Forestry Management in Germany', in Tore Fränsgmyr, J.L. Heilbron and Robin E. Rider (eds.) *The Quantifying Spirit in the 18th Century.* (Berkley: University of California Press) pp. 315–42.

Ma Lin, Feng Ri Li and Wei Wei Jia. 2011. 'An Analysis of the Carbon Content Rate of Natural White Birch in Xiaoxinganling Mountain Area'. *Advanced Materials Research* **393–5**: 580–6.

MacKenzie, J. M. 1988. *The Empire of Nature: Hunting, Conservation and British Imperialism.* (Manchester University Press).

Mahanty, Sanghamitra. 2002. 'Conservation and Development Interventions as Networks: The Case of the India Ecodevelopment Project, Karnataka'. *World Development* **30**(8): 1369–86.

Mantel, K. 1980. *Forstgeschichte des 16. Jahrhunderts unter dem Einfluss der Forstordnungen und Noe Meurers* [Forest History of the Sixteenth Century Under the Influence of Forestry Regulations]. (Paul Parey: Hamburg).

Mantel, K. 1990. *Wald und Forst in der Geschichte* [History of Forests and Forestry]. (Alfeld–Hannover: Verlag M. & Schaper, H).

Markus, Rudlofs. 1967. *Ostwald's Relative Forest Rent Theory.* (Munich: Bayerischer Landwirtschaftsverlag).

Martin, Adam R. and Sean C. Thomas. 2011. 'A Reassessment of Carbon Content in Tropical Trees'. *PLoS ONE* **6**(8) e23533: 1–9.*

Matérn, Bertil. 1960. 'Spatial Variation. Stochastic models and their application to some problems in forest surveys and other sampling investigations'. *Meddelanden fran Statens Skogsforskningsinstitut* **49**: 5.

McCracken, Donal P. 1997. *Gardens of Empire: Botanical Institutions of the Victorian British Empire.* (London: Leicester University Press).

McDermott, Constance L., Benjamin Cashore and Peter Kanowski. 2010. *Global Environmental Forest Policies: an International Comparison.* (London: Earthscan).

Menzies, Nicholas. 1988. 'Three Hundred Years of Taungya: A Sustainable System of Forestry in South China'. *Human Ecology* **16**(4): 361– 76.

Millennium Ecosystem Assessment. 2005a. *Ecosystems and Human Well-being: General Synthesis.* (Washington, D.C.: World Resources Institute).*

Millennium Ecosystem Assessment. 2005b. *Ecosystems and Human Well-being: Biodiversity Synthesis.* (Washington, D.C.: World Resources Institute).*

Mills, Jenny. 2002. 'Stoate, Theodore Norman (1895–1979)'. *Australian Dictionary of Biography*, v.16 (Melbourne: Melbourne University Press).*

Mitchell, S.J. and W.J. Beese. 2002. 'The retention system: reconciling variable retention with the principles of silviculture'. *The Forestry Chronicle* **78**(3): 397–403.*

Möhring, Bernhard. 2001. 'The German Struggle Between the "Bodenreinertragslehre" (Land Rent Theory) and "Waldreinertragslehre" (Theory of the Highest Revenue) Belongs to the Past – but What is Left?' *Forest Policy and Economics* **2**: 195–201.

Morris, Deirdre. 1974. 'Mueller, Sir Ferdinand Jakob Heinrich von [Baron von Mueller] (1825–1896)'. *Australian Dictionary of Biography* (Melbourne University Press) v.5. pp. 306–8.*

Nicolson, Dan H. 1991. 'A History of Botanical Nomenclature'. *Annals of the Missouri Botanical Garden* **78**(1): 33–56.

North Carolina State University Cooperative Tree Breeding Program. 2011. 55th *Annual Report*.*

Oosthoek, K.J.W. (forthcoming). *Conquering the Highlands: A History of the Afforestation of the Scottish Uplands* (Canberra: ANUePress).*

Paar M., I. Oberleitner and H. Kutzenberger. 1998. *Fachliche Grundlagen zur Umsetzung der Fauna-Flora-Habitat-Richtlinie* [Professional Elements for the Implementation of the Flora-Fauna-Habitat-Guideline]. (Wien: Bundesministerium für Umwelt, Jugend und Familie].

Parrotta, J.A. and R.L. Trosper (eds.) 2012. *Traditional Forest-Related Knowledge: Sustaining Communities, Ecosystems and Biocultural Diversity*. (Dordrecht: Springer).

Parry, M.L., O.F. Canziani, J.P. Palutikof, P.J. van der Linden and C.E. Hanson (eds.) 2007. *Climate Change 2007: Impacts, Adaptation and Vulnerability: Contribution of Working Group II to the Fourth Assessment Report of the Intergovernmental Panel on Climate Change.* (Cambridge University Press).*

Parviainen, Jari, Declan Little, Marie Doyle, Aileen O'Sullivan, Minna Kettunen and Minna Korhonen (eds.) 1999. *Research in Forest Reserves and Natural Forests in European Countries.* (Joensuu: European Forest Institute, EFI Proceedings No. 16).

Pearse, Peter H. 1990. *Introduction to Forest Economics.* (Vancouver: University of British Columbia Press).

Perlin, John. 1989. *A Forest Journey: The Role of Wood in the Development of Civilization.* (Norton: New York).

Peterson, Charles E. and Paul D. Anderson. 2009. 'Large-scale Interdisciplinary Experiments Inform Current and Future Forestry Management Options in the U.S. Pacific Northwest'. *Forest Ecology and Management* **258**: 409–14.

Petrini, Sven. 1928. 'A Bibliography of Recent Forest Literature in Sweden'. *Forestry* **2**: 110–25.

Pezzey, John C. V. and Michael A. Toman. 2002. *The Economics of Sustainability: A Review of Journal Articles.* (Washington D.C: Resources for the Future).*

Piggott, Cheryl. 2004. 'David Douglas (1799–1834)'. *Oxford Dictionary of National Biography.* (Oxford University Press)

Pockberger, J. 1952. *Die naturgemässe Waldwirtschaft als Idee und Waldgesinnung* [Nature-based Silviculture. Idea and Perception]. (Vienna: Fromme)

Poole, L.A. 1961. *New Zealand Forestry*. (Wellington: Govt. Printer).

Prain, David. 2004. 'Brandis, Sir Dietrich (1824–1907)'. Revised M. Rangarajan. *Oxford Dictionary of National Biography* (Oxford University Press).

Pravat, Poshendra Satyal. 2010. 'Looking Back to Move Forward: Using History to Understand the Consensual Forest Management Model in the Terai, Nepal'. *Global Environment* 6: 96–121.

Pressler, M.R. [1860] (trans. W. Lowestein and J.R Wirkner). 1995. 'For the Comprehension of Net Revenue Silviculture and the Management Objectives Derived Thereof'. *Journal of Forest Economics* 1: 45–87.

Pressler, Maximilian Robert. 1858, 1859. *Der rationelle Waldwirth und sein Waldbau des höchsten Ertrags* [The Rational Foresterand and his Silviculture of the Highest Yield]. (Dresden: Türk, 2 vols).

Pretzsch, Hans. 2009. *Forest Dynamics, Growth and Yield: from Measurement to Model.* (Dordrecht: Springer).

Proctor, Wendy. 2009. *Environmental Decision Making: An Application of Multi-criteria Analysis to a Case Study of Australia's Forests.* (Saarbrücken: Lambert Academic Publishing).

Puettmann, Klaus J., K. David Coates and Christian Messier. 2009. *A Critique of Silviculture: Managing for Complexity.* (Washington: Island Press).

Rackham, Oliver. 1996 [1976]. *Trees and Woodland in the British Landscape: the Complete History of Britain's Trees, Woods & Hedgerows.* (London: Phoenix).

Rajan, Ravi. 2006. *Modernizing Nature: Forestry and Imperial Eco-development, 1800– 1950.* (Oxford: Clarendon Press).

Raudiere, Pierre de la. 1930. 'De L'application des Engrais Chimiques aux Peuplements Forestiers'. *Bulletin du Comité des Forêts*: 180–197.

Rawat, Ajay S. 1993. 'Brandis: The Father of organized Forestry in India', in Ajay S. Rawat (ed.) *Indian Forestry: A Perspective.* (New Delhi: Indus Publishing Co.).

Recknagel, A.B., John Bentley and C.H. Guise. 1926. *Forest Management.* (New York: John Wiley) 2nd edn.

Reed, J.L. 1954. *Forests of France.* (London: Faber & Faber).

Richards, Paul W. 1952 [2nd edn. 1957]. *The Tropical Rain Forest: an Ecological Study.* (Cambridge University Press).

Samuelson, P.A. [1976] 1995. 'Economics of Forestry in an Evolving Society'. *Journal of Forest Economics* 1: 115–49.

Salisch, Heinrich von. 2009 [1885}. *Forstästhetik.* [Forest Aesthetics]. (Remagen: Kessel, 4th edn. based 3rd edn. 1911).

Sasaki, Nophea and Francis E. Putz. 2009. 'Critical need for new definitions of "forest" and "forest degradation" in global climate change agreements'. *Conservation Letters* 2: 226–32.

Schimper, A.F.W. [1898], revised F.C. von Faber. 1935. *Plfanzengeographie auf physiologischer Grundlage.* (Jena: Fischer).

Schimper, A.F.W., (trans. William R. Fisher). 1903. *Plant-geography on a Physiological Basis* (Oxford: Clarendon Press).

Schlaepfer, Rodolphe (ed.) 1993. *Long-term Implications of Climate Change and Air Pollution on Forest Ecosystems: Progress report of the IUFRO Task Force 'Forest, Climate Change and Air Pollution'.* (Vienna: IUFRO).

Schleicher, Francis Grant. 1917. 'Reclamation of Coastal Sand Dunes in Europe and in Eastern United States' (Ithaca, NY, Cornell University, mss).*

Schlich, W. 1889–1896, *A Manual of Forestry.* (London: Bradbury Agnew 5 vols).

Schlich, W. 1895. *A Manual of Forestry, Vol III Forest Management including Mensuration and Valuation.* (London: Bradbury Agnew).

Schneider auf Negelsfuerst, Freiherr von 1798. *Aufforderung zur allgemeinen Anpflanzung des unechten Akazienbaumes durch deren Anzucht dem Holzmangel bald dauerhaft abgeholfen werden könnte* [Call for the General Planting of False Acacia to Soon Permanently Remedy the Wood Shortage]. (located in FAO Rome, Lubin Library).

Schoene, Dieter, Wulf Killmann, Heiner von Lüpke and Mette Loyche Wilkie. 2007. *Definitional Issues Related to Reducing Emissions from Deforestation in Developing Countries.* (Rome: FAO, Forests and Climate Change Working Paper 5).*

Schwappach A. 1911. 'Die Versuche des Sonderausschusses der Deutschen Landwirtschafts-Gesellschaft für Forstdüngung in Neumannswalde bei Neudamm und auf den städtischen Riesefeldern bei Berlin' [The Trials of the Special Committee of the German Agricultural Society for Forest Fertilisation in the Woods near Neumann and Large Urban Area near Berlin]. *Deutsche Forstzeitung* **26**: 1061–68.

Scott, James C. 1985. *Weapons of the Weak: Everyday Forms of Peasant Resistance.* (New Haven: Yale University Press).

Scott, James C. 1989. 'Everyday Forms of Resistance', in Forest D. Colburn (ed.) *Everyday Forms of Peasant Resistance* (Armonk, New York: M.E. Sharpe), pp. 3–33. Also published in *Copenhagen Papers* **4**: 33–62.*

Scott, Sir Peter, John A. Burton and Richard Fitter. 1987. 'Red Data Books: the Historical Background', in Richard and Maisie Fitter (eds.) *The Road to Extinction; Problems of Categorizing the Status of Taxa Threatened with Extinction.* (Gland, Switzerland: IUCN), pp. 1–5.

Shirley, James W. 2008. *A History of Australian Capital Territory Arboreta, 1928–2003.* (Melbourne: Forestry and Wood Products Australia).*

Sieferle, Rolf Peter (trans. Michael P. Osman). 2001. *The Subterranean Forest: Energy Systems and the Industrial Revolution* (Cambridge: White Horse Press), first published in German in 1982 as *Der Unterirische Wald* (Munchen: C.H. Beck).

Singh, Preet Pal. 2008. 'Exploring Biodiversity and Climate Change Benefits of Community-based Forest Management'. *Global Environmental Change* **18**: 468–78.

Smout, T.C., Alan R. MacDonald and Fiona Watson. 2005. *A History of the Native Woodlands of Scotland.* (Edinburgh University Press).

Soegaard, Bent. 1968. 'C. Syrach-Larsen, Pioneer in Forest Tree Breeding'. *Silvae Genetica* **17**(5–6): 157–8.

Spurr, Stephen H. 1956. 'German Silivicultural Systems'. *Forest Science* **2**(1): 75–80.

Steffen, Will, Andrew A. Burbidge, Lesley Hughes, Roger Kitching, David Lindenmayer, Warren Musgrave, Mark Stafford-Smith and Patricia A. Werner. 2009. *Australia's Biodiversity and Climate Change.* (Collingwood, Victoria: CSIRO Publishing).

Stoate, T. N. and C.E. Lane Poole. 1938. *Application of Statistical Methods to Some Australian Forest Problems, vol. 1.* (Commonwealth Forestry Bureau, Australia, Bulletin 21).

Stone, Lois C. 1968. 'A Legacy of Good Breeding'. *Forest History* **12**(3): 20–29.

Sunseri, Thaddeus. 2009. *Wielding the Axe: State Forestry and Social Control in Tanzania, 1820– 2000.* (Athens, Ohio: Ohio University Press).

Syrach-Larsen, C. (trans. M.L. Anderson). 1956. *Genetics in Silviculture.* (Edinburgh: Oliver and Boyd).

Tamm, C.O. 1968. 'The Evolution of Forest Fertilization in European Silviculture', in *Forest Fertilization, Theory and Practice.* (Muscle Shoals, Alabama: Tennessee Valley Authority) pp. 242–7.

Taylor, Benedict. 2008. '"Trees of Gold" and Men Made Good? Grand Visions and Early Experiments in Penal Forestry on the NSW North Coast, c.1913–1938'. *Environment and History* **14**: 545–62.

Temple, Samuel. 2011. 'Forestation and its Discontents: The Invention of an Uncertain Landscape in Southwestern France, 1850–Present'. *Environment and History* **17**: 13–34.

Thomasius, H. 1992. 'Prinzipien eines ökologisch orientierten Waldbaus' [Principles of Ecology-oriented Silviculture]. *Forstwissenschaftliches Centralblatt* **111**: 141-155.

Trosper, R.L. and J.A. Parrotta. 2012. 'Introduction: The Growing Importance of Traditional Forest-Related Knowledge', in J.A. Parrotta and R.L. Trosper (eds.) *Traditional Forest-Related Knowledge: Sustaining Communities, Ecosystems and Biocultural Diversity.* (Dordrecht: Springer).

Trosper, Ronald L., John A. Parotta, Mauro Agnoletti, Vladimir Bocharnikov, Suzanne A. Feary, Mónica Gabay, Christain Gamborg, Jósus García Latorre, Elisabeth Johann, Andrey Laletin, Lim Hin Fui, Alfred Oteng-Yeboah, Miguel Pinedo-Vasquez, P.S. Ramakrishnan and Youn Yeo-Chang. 2012. 'The Unique Character of Traditional Forest-Related Knowledge: Threats and Challenges Ahead', in J.A. Parrotta and R.L. Trosper (eds.) *Traditional Forest-Related Knowledge: Sustaining Communities, Ecosystems and Biocultural Diversity.* (Dordrecht: Springer).

Troup, R. S. 1928. *Silvicultural Systems.* (Oxford: Clarendon Press).

Troup, R.S. 1921. *The Silviculture of Indian Trees.* (Oxford: Clarendon Press, 3 vols).

Tsouvalis, Judith and Charles Watkins. 2000. 'Imagining and Creating Forests in Britain, 1890–1939', in M. Agnoletti, and S. Anderson (eds.) *Forest History: International Studies on Socio–Economic and Forest Ecosystem Change.* (Wallingford: CABI Publishing).

Turckheim, Brice de. 2006. 'Economic Aspects of Irregular, Continuous and Close to Nature Silviculture ((SICPN): Examples about Private Forests in France', in Jurij Diaci (ed.) *Nature-based Forestry in Central Europe: Alternatives to Industrial Forestry and Strict Preservation* (Ljubljana: University of Ljubljana. Studia Forestalia Slovenica 126).

United Nations Conference on Sustainable Development. 2012. *The Future We Want*: [Outcome Document of Rio+20, Earth Summit 2012 Conference] (UNCSD).*

USDA, United States Department of Agriculture, Forest Service. 1987. *Kootenai National Forest Plan: Record of Decision.**

Vavra, Elisabeth (ed.) 2008. 'Der Wald im Mittelalter. Funktion – Nutzung – Deutung. [Forest and Woodland in the Middle Ages. Function – Utilization – Interpretation]'. *Das Mittelalter* **13**(2): 197 pp.

Vilmorin, M. [Philippe André] de. 1864. *Exposé Historique et Descriptif de L'École Forestière des Barres (Loiret)* [History and Description of the Forestry School at Barres (Loiret)]. (Paris: Buchard-Huzard).*

Ware, Glenn O. and Jerome L. Clutter. 1971. 'A Mathematical Programming System for the Management of Industrial Forests'. *Forest Science* **17**(4): 428–45.

Waterworth, R.M. and G.P. Richards. 2008. 'Implementing Australian Forest Management Practices into a Full Carbon Accounting Model'. *Forest Ecology and Management* **255**: 2434–43.

Watson, Robert T., Ian R. Noble, Bert Bolin, N.H. Ravindranath, David J. Verado and David J. Dokken (eds.) 2000. *Land Use, Land-Use Change, and Forestry: IPCC Special Report of the Intergovernmental Panel on Climate Change* (Cambridge University Press).

Watts, Kevin. 2006. 'British Forest Landscapes: the Legacy of Woodland Fragmentation'. *Quarterly Journal of Forestry* **100**(4): 273–9.*

Westoby, Jack. 1987. 'Forest Industries for Socio-Economic Development', in Jack Westoby with Foreword by A.J. Leslie. *The Purpose of Forests: the Follies of Development.* (Oxford: Basil Blackwell).

Westoby, Jack C. 1962. 'Forest Industries in the Attack on Underdevelopment'. *Unasylva* **16**(4), no. 67: 168-201.*

White, Andy and Alejandra Martin. 2002. *Who Owns the World's Forests? Forest Tenure and Public Forests in Transition.* (Washington, DC: Forest Trends and Center for International Law).*

White, Tim L. and Tom D. Byram. 2004. 'Slash Pine Tree Improvement', in E.D. Dickens, J.P. Barnett, W.G. Hubbard and E. Jokela (eds.) *Slash Pine: Still Growing and Growing!* (Asheville, N.C: US Dept of Agriculture, Forest Service, Southern Research Station).*

White, Timothy L., W. Thomas Adams and David B. Neale. 2007. *Forest Genetics.* (Wallingford: CABI Publishing).

Wiedemann, E. 1925a: [1st edn. 1923]. *Zuwachsrückgang und Wuchsstockungen der Fichte* [Regression of Increment and Stagnation of Growth of Spruce]. (Tharandt: W. Laux) 2nd edn.

Wiedemann, E. 1925b. Fichtenwachstum und Humuszustand [Spruce Growth and Humus Condition] *Arbeiten aus der Biologischen Reichsanstalt für Land- und Forstwirtschaft* **13**: 1–27.

Williams, Michael. 2003. *Deforesting the Earth: From Prehistory to Global Crisis.* (University of Chicago Press).

Williams, Michael. [1989] 1999. *Americans and Their Forests: a Historical Geography.* (Cambridge University Press).

Willis, Kathy J. and Shoonil A. Bhagwat. 2009. 'Biodiversity and Climate Change'. *Science* **326**: 806–7.

Wobst, Hermann. 2006. 'Combination of Economic and Ecological Aspects by Close to Nature Forestry: A Contribution to the Economic Crisis of Forestry', in Jurij Diaci (ed.) *Nature-based Forestry in Central Europe: Alternatives to Industrial Forestry and Strict Preservation* (Ljubljana: University of Ljubljana. Studia Forestalia Slovenica 126).

World Commission on Environment and Development [Brundtland Commission]. 1987. *Our Common Future.* (Oxford: Oxford University Press).

Wu, Harry X., Ken G. Eldridge, A. Colin Matheson, Mike B. Powell, Tony A. McRae, Trevor B. Butcher and Ian G. Johnson. 2007. 'Successful Introduction and Breeding of Radiata Pine in Australia'. *Australian Forestry* **70**(4): 215–25.*

Ximenes, F.A. and W.D. Gardner. 2006. *The Decay of Coarse Woody Roots Following Harvest in a Range of Forest Types.* (Canberra: Department of Environment and Heritage, Australian Greenhouse Office, Technical Report 49).*

Young, Harold E. and Paul M. Carpenter. 1964. *Weight, Nutrient Element and Productivity Studies of Seedlings and Sapplings of Eight Tree Species in Natural Ecosystems.* (Orono, Maine: University of Maine, Technical Bulletin 28).*

Zacharin, Robert Fyfe. 1978. *Emigrant Eucalypts, Gum Trees as Exotics.* (Melbourne University Press).

Zobel, Bruce and John Talbert. 1984. *Applied Forest Tree Improvement.* (New York: John Wiley).

Zobel, Bruce J. and Jerry R. Sprague. 1993. *A Forestry Revolution: The History of Tree Improvement in the Southern United States.* (Durham, N.C.: Forest History Society).

INDEX

A

aboriculture 4, 65, 68, 87, 113, 193
aboriginal people 195–196
acacia 71, 120, 124
  see also common acacia
Academy of Sciences, French 6, 28, 110
acid rain 130
adaptation to climate change 210–211, 215–216
adaptive management 140
aerial photography 16
Africa
  colonies 31
  deforestation 173
  East 170
  North 119, 178
  Sub-Saharan 148
  West 74, 94, 118
agriculture 82, 124, 196
  conservation 170, 179, 223–224
  land 63, 91, 98, 126, 188, 209
  see also Slash and burn
agronomists 193
air pollution 112, 130, 208
Allgemeine Forst- und Jagzeitung 12
allowable cut 55, 57, 60
alpine
  farming landscapes 223–224
  forests 29, 77, 113, 162, 164, 169
  protected areas 180
Amazon *see* Forest of
Anderson, Mark Louden (1895-1961) 88–91, 115
anthropologists 185, 189, 192, 193
Arbeitsgemeinschaft Naturnahe Waldwirtschaft (Working Party for Close to Nature Forestry) 164
arboreta 65, 72–73, 113, 115
  Des Barres 111–112
  Duhamel's 110
  Hørsholm 115
  pinetum 68
arboricultural Societies 68, 87
arboriculture 4, 65, 113, 193
Arrhenius, Svante August (1859–1927) 203
ash 58, 70, 89
assessment *see* measurement, and survey
Australia 38, 112, 114, 116. 118, 119, 120, 135, 158
  carbon model 206
  Codes of Forest Practice 151
  conservation reserves 178–179
  growth and yield models 136
  nutrient disorders 126
Austria 11–12, 37, 41, 77, 112–113, 125, 152, 223
  conservation reserves 168–169, 180
  fertiliser trials 125
  rotation 47
Austrian method 55
Austrian pine 78, 89

B

Baden 12
Baker, Francis Storrs (1890–1965) 31
Bangladesh 94
Banská Štiavnica 12
basal area 18, 21–22
basic slag 125
Bavaria 12, 55, 77
beauty 5, 40, 48, 73, 149, 168–169
beech 34,35, 37, 58, 77, 89, 125
Belgium 86, 93, 125
bioclimatic envelopes 214
biodiversity 7, 60, 151, 164, 168, 171, 173–174, 178, 180, 196, 223
  and cultural diversity 222
  climate change 204, 206, 208, 209, 213–216
  indigenous and local people 194–196
  sustainable 219–220
biogeography 213
Biolley, Henri (1858–1939) 58–59, 161
biomass 205, 207, 212, 215

biosphere 203
biostatistics 38
Bitterlich, Walter (1908–2008) 21–22
black pine 82, 125
Bohemia 77
botanical gardens 65, 71, 96
  Adelaide 73
  Berlin 65
  Denmark 115
  Kew 96, 115
  Le Jardin des Plantes 65, 110
  Schönbrunn 65
  St Petersburg 65
botanists 65–69, 71–72, 93, 96, 99, 156, 193, 210, 213–214, 230
Brandenburg 77
Brandis, Lady Katinka 95, 100
Brandis, Sir Dietrich (1824–1907) 59, 94, 96, 99–103
Brazil 63, 71, 74, 117–120, 142, 232
breeding 74, 108, Ch.8
  gains 120
  protocol 116–117
Bremontier, Nicolas Thomas (1738–1809) 83
Britain 5–6, 11, 12, 29, 31, 37, 68, 70, 74, 87, 91, 135, 230
British East India Company 93
British India 94–104
broad-leaved trees *see* hardwoods
Brown, John Ednie (1848–1899) 73
Buffon, Georges-Louis Leclerc Comte de (1707–1788) 6, 110
Burgsdorf, Friedrich August von (1747–1802) 111
Burma 58, 59, 93–94, 99, 100, 103
Burundi 170

C

calcium 125, 142
California 69, 70, 71, 112, 113, 121, 214
cameralism 12, 24, 29, 40, 54, 231
Canada 31, 69, 107, 112, 128, 135, 187
Candolle, Alphonse Louis Pierre Pyrame de (1806–1893) 66
canopy 34, 36, 161
carbon 123, 203–209
  see also Measurement
carbon dioxide 130, 203, 210–213, 216
  see also measurement
Carl, Count of Hessen 111
Carlowitz, Hans Carl von (1645–1714) 111
Cato the Elder (234–149 BC) 5
Centre for International Forestry Research 193
Chambrelent, Jules (1817–1893) 83
Chapman, Herman Haupt (1874–1963) 57
charcoal 11, 41, 67, 142
chestnut 2, 27, 81, 89
Chevandier [de Valdrôme], Eugène (1810–1878) 125
Chile 63, 74, 120
China 66, 72, 103, 118
Chipko movement 186–187
chir pine 188–189
Christmas trees 128
Ciéslar, Adolf 1858–1934) 112
clear-cut areas 56, 152
clear-cutting 27, 29, 33–34, 47, 53, 77, 150, 164
  critique 159–161
clear-felling *see* clear-cutting
clearing *see* deforestation
Cleghorn, Hugh (1820–1895) 100
climate change 7, 60, 71, 120, 131, Ch 14; conferences 204
climatic zones 70, 97
  see also homoclimes
clones 26, 114, 115, 118, 119, 144
close-to-nature forestry *see* nature-based forestry
Clutter, Jerome L. 'Jerry' (1934–1983) 135–136, 140
coal 11, 29, 41, 87
  emissions 203
coastal dunes, stabilisation 83, 86
Codes of Forest Practice 151–152, 159, 215
Colbert, Jean-Baptiste (1619–1683) 5–6, 28
collaborative forest management *see* social forestry
common acacia 67–68

Index

commons 6, 81–82, 84–85, 184, 191, 193, 223
community
  forest 161–162
  scientists 219
  social expectations 192, 194, 196–197
computers 15, 24, 38, 107, 134–135, 138, 140–141, 156, 172–173, 203
concessions, timber 104
Congo 94, 119
conifers 6, 25, 32, 38, 41–42, 47, 68–69, 71, 81–82, 91, 111, 113, 124–125, 172, 210, 231
conservation biology 204
conservation refugees 194
  see also eviction
conservation reserves 168–171
  design 178–180
  Indonesia 170
  Slătioara Forest 169
  Württemberg 169
  see also protected areas
conservationist's vision 168–170
consultation *see* public participation
continuous forest cover 163, 164
Continuous Forest Inventory 24, 142
control method 16, 24, 58, 149, 162
Convention on Biological Diversity 178, 221
Convention on International Trade in Endangered Species of Wild Fauna and Flora 192
cooperatives 107, 142
  fertiliser 128, 129
  planning 136, 140–141
  breeding 116, 216
coppice 1, 27–28, 52
Cotta, Johann Heinrich (1763–1844) 12
coupe 28, 33, 159, 161
Crescentiis, Petrus de (1230/35–ca.1320) 2
Croatia 78
Crown Zellerbach 128
Cunninghamia 103, 118
cuttings 114, 118, 119
Czech Republic 130–131

D

Dalhousie, James Andrew Broun Ramsay, tenth Earl and first Marquess of (1812–1860) 99
Darwin, Charles Robert (1809–1882) 109, 213
debt *see* interest rate
Decades of Development 184, 201
  First 133
  Second 148
Declaration on the Rights of Indigenous Peoples 192, 221
deep ecology 158
deforestation 1, 3, 76–78, 171–174, 186, 201, 221, 206, 215
deglupta 121
degradation 78, 121, 173–174, 186, 201, 221, 215
Dehradun 31, 100
demand *see* wood demand
demarcation 16, 98, 103
Denmark 77, 86, 114
desiccation 203
developed countries 107, 185–186, 196–197
developing countries 107, 133, 151, 184–186, 193–194, 196–197
development 92, 107, 133, 144, 147–148, 183–188, 193, 216
  see also Decades of Development, Sustainable Development and wood demand
discount rate *see* interest rate
discounted cash flow 44, 140
discounting 44, 49, 225
dispossession *see* eviction
distributions *see* statistical analysis
diversity, cultural and biological 222
  see also biodiversity
DNA 110
Douglas fir 56, 68–69, 70, 77, 115, 128
Douglas, David (1799–1834) 69–70, 115
drainage of moors 82–89
Duhamel du Monceau, Henri-Louis (1700–1782) 6, 28, 110
Dutch East India Company 93

E

Earth Summit *see* UN Conference 1992
earth system models 215
eastern white pine 68, 70, 77
ecodevelopment 194
ecologists 168, 194–195
  botanical 174–175
  vision 168–170, 180, 183
  wildlife 156. 175–176, 194
ecology 34, 76, 88, 90–91, 123, 150, 152, 158–165, 168–169, 171–174, 188
economics 6, 12, Ch 3, 60, 70, 117, 130–131, 133–134, 137, 153–157, 164, 204, 210, 215
  ecological 224
ecosystem services 149
ecosystems 27, 130, 149, 159, 164, 174, 178–179, 201, 205, 230
Eddy, James Garfield (1881–1964) 113
elephants 93
elm 3, 70, 81, 89, 110
empires 93, 107
England *see* Britain
Enlightenment 11, 40, 54, 68, 219
environmental era 50, 149–150, 154, 156, 161, 168, 171, 180, 183, 186, 192, 204, 219, 224, 233
environmental gradient 159
environmental impact statements 147, 157, 183–184
environmentalism 147
epistemology 188–189
erosion 29, 36, 48, 152, 159, 171, 188
eucalypts 70–74, 102, 121, 188
  forests 28,159, 212
  plantations 38, 63, 72, 118–120, 142, 232
Europe 2–3, 6, 7, 18–19, 27, 34, 38, 41, 52, 65– 66, 111–112, 131, 135–136, 149, 164–165, 168–170, 173–174, 179, 192, 196, 205, 210, 213, 223, 231
Europe, Central 341, 41, 47–48, 52, 91, 130, 144, 161, 164
Europe, western and southern 77
European Union 179, 197, 205
Evelyn, John (1620–1706) 3–5, 40, 53, 110, 230
even-aged *see* forests, and silviculture
eviction 98, 103, 169, 170, 176, 194–195, 222, 232
exhibitions, international 84
experiment stations 12–13, 31
exports *see* wood exports
extinction *see* species, and wildlife

F

false acacia *see* common acacia
FAO 107, 133, 171–174, 184–185
  Codes of Forest Practice, 151
  community forestry 189
farm forestry *see* social forestry
fauna *see* wildlife
Faustmann, Martin (1822–1876) 44
Faustmann/Pressler economic model 47–48, 50, 224
Federation of Community Forest Users 189
fertilisers 38, 50, 87, 91, 108, Ch 9, 133, 135–136, 138, 142–144, 205, 215–216, 232, aerial 128
fertility, soil 88–89, 123
Fielding, John Mervyn (1910–1995) 119
Finland 32, 128, 169, 179
fir *see* silver fir
fire 19, 27, 32, 61, 85, 138–139, 155, 159, 173, 179, 195–196, 209, 215
firewood 86, 187
  see also fuel wood
Fisher, Sir Ronald Alymer (1890–1862) 109, 126
Flora and Fauna Preservation Society 170
fluxes, fluxnet 207, 211–213
foliar analysis 130
Food and Agricultural Organisation *see* FAO
forest
  Adirondack 57
  African woodlands 184
  Amazon 174, 184, 220
  Bago 212
  Black Forest 129

Index

Boubín virgin forests 169
Couvet 162
Douglas fir region 56
Kootenai 157
Les Landes de Gascogne 85
Nepal 188–190
Neuchâtel 58
Niger 187–188
Otways 156,158
Regensberg 35, 52, 163
Sundarbans 176
Tronçais 52
Vimperk 168
Vosges Mountains 58
Willamette 52
Zátoň 169
forest definition 6, 171
Forest Research Institute
Austria 125
India 31, 100, 102
Malaya 104
Forest Science 24
Forest Service
India 94, 100, 190
Niger 187
US 49,150, 155–156
forest type 96, 174
boreal 207–208
climax 173
closed forest 171
monsoon 97
old-growth 169, 173
open forest 172
primaeval 173
primary 173–174
savannah 97, 206, 213
secondary 173
temperate 207–208
tropical 207–208
tropical savanna 207–208
virgin 169, 173
woodland 172
see also coppice
forest-dependent people 98, 186, 192–195
foresters 11–12, 25, 63, 94, 96
cohesion 230
imperial, 192
militaristic 52
vision 27, 40, 47–49, 93, 107, 150, 168, 180, 183
forestry schools 12, 31, 82, 94, 96, 118
Austria 80
Australia 127
France 72, 94
Germany 54, 77, 94
see also university
form factor 18, 21
formations 97
fossil fuels 203
France 11, 31, 37, 47, 53, 58, 93, 111, 125
fuel wood 11, 27–28, 41, 57, 98, 148, 188, 186, 223
functions 149–150

G

game parks 170
Ganges delta 176
Gayer, Karl (1822–1907) 29–30, 34
genetic diversity 117
genotype 109–110
geographic information systems 142, 144, 158, 214
German Agricultural Society 125
German Association of Forest Research Stations 125
Germany 11, 31, 37, 77, 86, 111
empire 93, 170
global positioning systems (GPS) 21
globalisation 183, 202, 221
goats 78, 188, 223
Goricia 78
grafting 114
grazing 27, 40, 60, 77–78, 82, 84, 124, 149, 153, 168, 186, 191, 223
green energy 205
greenhouse effect 203
growing stock 49, 55, 59
guano 123
Gurnaud, Adolphe (1825–1898) 58

H

habitat trees 161
hardwoods 25, 27, 32, 38, 41, 47, 76, 81–82, 96, 111, 124–125, 144, 164, 172, 189
Hartig, Georg Ludwig (1764–1837) 12, 54
Havelock, General Sir Henry (1795–1857) 100
hazel 27
health 130–131, 151, 208
herbaria 96–97, 179, 214
heritage
  value 184, 223
  World 176
Heyer, Carl Justus (1797–1856) 55, 57
high-grading *see Jardinage*
Hiley, Wilfred Edward (1886–1961) 48–49
Himalayas 96, 99, 186
homoclimes 72, 203, 214
Hooker, W.J. (1785–1865) 69
hormones 119
hornbeam 58
Hufnagl, Leopold (1857–1942) 57
humus 78,125, 207–208
Hundeshagen, Johann Christian (1783–1834) 43, 53–55, 57
Hungary 11
hunting 12, 54, 60, 98, 149, 168, 170, 197, 223
Hutchins, David Ernest (1850–1920) 72
hybrids 113–114, 118, 120

I

ilimo 97
Imperial Forest College (India) 100
Imperial Forestry Institute, Oxford 127
in-breeding 117–118
increment 23, 41, 43, 47, 55, 57, 59, 101
India 31, 71–72, 74, 93–104, 118, 169, 176, 186–187, 190–192, 194–195, 215
Indian Engineering College 94
Indian Forester 101
indigenous culture, knowledge and language 2, 192, 193, 219–222
indigenous management 191, 195
  see also social forestry
indigenous people 169–170, 183, 186, 192–194, 196–197
  see also eviction
Indonesia 31, 209
  conservation reserves 170, 176
  deforestation 173
  see also development
industrialisation 11, 40, 82, 131, 133, 148, 184, 216
Institute of Forest Genetics, USA 113
interest rate 44, 46–49, 50, 225
Intergovernmental Panel on Climate Change (IPCC), 204, 206–207, 211
international aid agencies 184, 187
International Biosphere Reserve 176
International Conference on African Wildlife 179
International Council for Research in Agroforestry 193
International Development Research Centre (Canada) 187
International Paper Company 142–144
International Phenology Gardens 210
International Union for Nature Protection 170
International Union for the Conservation of Nature (IUCN) 147, 170, 175–179, 222
International Union of Forest Research Organisations (IUFRO) 13, 31, 37, 125, 219, 221
  Congress 112
inventory *see* measurement
Ireland 151
Italy 93

J

Japan 31
Japanese cedar 118
Jardinage 28, 59
Java 94, 103
John, Josef (1802–1871) 168
joint management 190–192
Journal of Forestry 24
juniper 78

Index

K

karst region 34, 78–82
kerosene 185
Kessell, Stephen Lackey (1897–1979) 127
knowledge
  indigenous and traditional 192–193, 219–221
  scientific 6, 219–224
  transfer 6, 12, 94, 134, 156
Kuhn, Thomas Samuel (1922–1996) 219

L

laissez faire 41, 84, 231
land expectation value 44
Land-rent 43
lands of new settlement 94
  see also Australia, New Zealand, North America, South America, indigneous people
Landsat *see* satellites
landscape 178, 223–224
Lane Poole, Charles Edward (1885–1970) 73, 96, 193
language, of science 25, 30, 31, 90, 94, 229, *see* indigenous language
larch 77, 78, 112, 120
League of Nations 201
legislation 11, 52, 85, 147, 168, 179, 184, 223
Leibundgut, Hans (1909–1993) 161–162
Leontif, Wassily (1905–1999) 224
Les Landes de Gascogne 78, 82–86
libertarianism 183
licence fees 98
Liebig, Justus von (1803–1873) 123
lime fertiliser 125, 131, 212
lime tree 58, 81, 89, 125
limestone *see* karst region
linden *see* lime tree
linear programming 140–141
Linnean Society 109
Liocourt, François de 58–59
loblolly pine 116, 129, 136
local communities 178, 186–197, 206

M

MacMillan Bloedel 127–128
macronutrients 123
Madagascar 118
Madras 100
Mahat, T.B.H. 189
mahogany 94
management 5, 6, 12, 16, 21, 40, 94, 98, 104, 151, 155, 169–170
  climate change 204, 206, 208, 215–216
  intensive planning Ch 10, 156–158
  multiple-use Ch 11
  objectives 48–49, 148, 168, 178
  protected areas 178–179
  sustainable 150–151, 165, 201, 218–220, 223–225
  see also nature-based forestry, regulation, social forestry
mangium 74, 121
Mantel, Wilhelm Wolfgang, von (1904–1983) 55
marine stores 86
maritime pine 83, 86, 91, 124–126
marram grass 83
Matérn, Bertil (1917–2007) 142–143
Mauritius 93
measurement
  age 18, 23, 99
  area 15–16
  biomass 205
  carbon dioxide, atmospheric 203
  carbon, forest stock 204–208
  deforestation 172–173
  diameter 16–17, 99
  fluxes 211–213
  forest health 208
  growth 23, 101
  height 17, 23, 99
  Relascope 21
  volume 18–24
  wildlife 175
Mecklenburg 77
Mendel, Gregor Johann (1822–1884) 109
mensuration *see* measurement
metabolic rift 123, 144
meteorologists 214

methane 130, 203
Mexico 31, 72, 214
micorrhiza 124
micro-climate 214
micronutrients 123
millennium goals 202
minor forest products 41, 193
mitigation, forest 215
Miwok people 169
Mlinšek, Dušan (1925–) 164
models
comprehensive planning 156–158
Earth system 215
global vegetation 215
growth and yield 38, 135–136, 138, 232
input/output 133, 224
optimising 140–141, 225
simulation 137–139
volume 18–24
modernisation 85, 184, 202
monocultures 32, 38, 41, 45, 77, 82, 90, 124
monsoon 99, 214
Morocco 119
mountain ash
Australian 32
European 81
Mueller, Sir Ferdinand Jakob Heinrich, Baron von (1825–1896) 71
multiple-use 150–158, 179, 195, 218, 220, 232
multiplier effects 185
Munro, A.V. 99, 101
Myanmar *see* Burma

N

Nancy *see* forestry schools, French
Näslund, Manfred Eugén (1899–1988) 143
National Park movement 168
see also protected areas, eviction
National Parks
Australia 158
Austria 180
Nepal 195
Finland 179
India 194
Switzerland 168
United Kingdom 168
USA 169
National Trust, UK 168
Natura 2000 179–180
Nature Conservation Parks Society 169
nature protection 168–170, 179
see also conservation reserves
nature-based forestry 34, 149, 151, 158, 161–165, 232–233
Nepal 188–190
net present value 44, 47, 141, 156
Netherlands 93, 125, 169
New World 29, 55
New Zealand 38, 63, 111–115, 120, 126, 135, 158, 187
niche 159
Niger 187–188
Nigeria 103, 104
nitrogen 123–125, 128–132, 142, 211
non-governmental organisations (NGOs) 171, 176, 187, 190, 194, 221–222
non-industrial species 120
non-timber forest products 193, 196
see also minor forest products
normal forest 44, 53–58, 60, 137, 151, 232
rejection 140
North America 18, 24, 38, 112, 128, 206, 221
trees 70, 111
Pacific Northwest 128
Norway 128, 130, 169
Norway spruce *see* spruce
nurseries 5, 70, 77, 81, 111, 124, 188–189
nutrient cycling 129, 205–208, 215
nutrients 91, 123–129

O

oak 3, 27, 43, 47, 58, 78, 81, 112
oil
emissions 203
price shock 147, 185
ontology 220
operations research 134–136, 154–155, 225
optimisation 140–144

Index

Oregon 112, 129
  Codes of Forest Practice 151
Ostwald, Eugen (1851–c.1931) 49
ozone 130–131

P

Pakistan 94
panchayats *see* village councils
Papua New Guinea 74, 96–97, 119, 121, 193, 222
paraffin 185
participation *see* public participation
participatory forestry social forestry
peasants 102–103, 124, 149, 184–188, 190, 218
peat 88–89, 203
Pegu 100–101
Pfeil, Friedrich Wilhelm (1783–1859) 12, 43
phenology 210
phenotype 109–110
Philippines 31, 121
phosphorous 123, 129, 142
plant collectors 65–66, 69–70, 84, 96, 110
plant community 88–89
plant nutrition 123
plantations 3, 6, 19, 36–38, 68–74, 85, 91, 118–121
  carbon, 205–206, 215
  critique or failure, 34, 72–73, 88, 91, 112, 126
  industrial, 7, 32, 38, 74, 82, 86, 109, 129–130, 133–144, 184
  social, 88–89, 191, 232
plantations of
  Brazil 120, 142–144
  Les Landes de Gascogne 82–85, 125
  Scotland, Britain 68–70, 88–91
  South Australia 73, 90, 127, 129
Pliny the Elder (AD 23–79) 2
plus trees 114
Poland 130–131, 152, 169
poles, wooden 27, 86, 186–187
pollen, preserved 76
pollination 112–116
ponderosa pine 112, 129
poplars 118–120
Portugal 63, 93, 118
positivism 219
potash 2, 125
potassium 123, 142
poverty 81–82, 148, 192–194, 201, 215
Pressler, Max Robert (1815–1886) 44–46, 50
productivity
  breeding 117
  climate change 204–213
  complete 205
  decline 129–130
  fertilisation 128
  wood 19, 55, 58, 61–63, 82, 92, 93, 107–108, 128
progeny trials 114, 116–117
Project Tiger 176, 194–195
propagation 118–119
property regimes 190
ProSilva 164
protected areas 170–171,178–180, 222
  see also conservation reserves
provenance 68, 70, 111–112, 118–119, 142
Prussia 12, 77, 100
Pryor, Lindsay Dixon (1915–1998) 119
public participation 147–148, 150, 156–158, 183, 189, 191, 197, 233
pulp and paper industry 116, 133, 137

Q

qualitative research 188

R

radiata pine 70, 71, 91, 116, 120, 126, 214
  declining yield 129–130
rainbow gum 121
rainforest 94, 97, 173, 186, 192, 208
  see also forest type
Rangoon 100
rate of return 46–47
recreation 150–151, 153–158, 164, 178, 196, 220
Red Book 175
red spruce 57

REDD and REDD+ 205–206, 216
reforest 67, 78, 81, 91, 187–188, 224
Regional Community Forestry Training Center for Asia and the Pacific 194
regression *see* statistical analysis
regulation 52–61, 135–142
  Brandis' method 101
  by area 53
  by diameter limit 99
  by volume 53
  even-aged forests 53–58
  formulas 55–57
  India 102
  limitations 60–61
  nature-based silviculture 161
  selection forests 58–59
  see also multiple-use
Relaskop-Technik 22
Reni village 186–187
resilience 27, 209, 214
resin 82, 86, 193
Ressel, Josef Ludwig Franz (1793–1857) 78–82
restoration 3, 6, 11, 27–28, 34, 55, 63, 67, 77, 87, 92, 111, 131, 159, 173, 215, 230, 232
  see also silviculture
Ribbentrop, Berthold Georg Theodor (1843–1917) 94, 100
Riga 49
river red gum 188
Robinson, Sir Roy Lister [Baron Robinson] (1883–1952) 87–88
rotation 43–50, 53, 55, 77, 134–139, 142, 162, 224
  second 129–130
Royal Indian Engineering College 100
Royal Society 3–6, 110
rural development forestry *see* social forestry
Russia 93, 111
rust resistance 117, 119
Rwanda 170

S

sal 95, 96
sampling 16, 19, 21–22, 24, 60
Samuelson, Paul Anthony (1915–2009) 48
São Paulo, State of 142
satellites 172, 204, 206, 210
satisficing 225
Saxony 12, 41, 77
Scandinavia 11, 114, 144
Schimper, Andreas Franz Wilhelm (1856–1902) 88, 96, 100
Schlich, Sir William (Wilhelm) Philipp Daniel (1840–1925) 48, 94, 96, 100
Schomburgk, Moritz Richard (1811–1891) 73
scions 114
Scotland 6, 32, 68–70, 87–91
Scots pine 41, 68, 77, 128
  provenances 111–112
seed 33, 81, 114, 118, 188
  introductions 65, 71–73, 111
  natural 27, 32–34, 101
  orchard 114–118, 188
  sowing 2, 27, 81, 83, 102–103
  see also provenance
selection *see* silviculture
shelter-wood 33, 34
shifting cultivators *see* slash and burn
Siberia 112
Sierra Leone 96
Silesia 77
Silvae Genetica 114
silver fir 34, 58, 70, 77, 89–90, 130
silviculturalists 31, 102, 104, 164
silviculture 2, 6, 45, 60, 110, 126, 131, 231, Ch.2
  critique 155, 158
  enrichment planting 103
  even-aged 32–34
  Indian 94, 96, 102–104
  selection 34–36, 58–59, 163
  shelter-wood 33–34
  taungya 103
  variable-retention 161
  see also nature-based forestry
Simla 100
simplification 24–25, 32, 50, 140, 159, 165, 220, 231–232
simulation 135–139, 156, 211, 215

site class, index, quality 19, 21, 23, 55, 81, 88–89, 140
slash and burn 103, 124
slash pine 116–120, 129, 136
Slovakia 12, 130–131
Slovenia 11, 16, 78, 152
social forestry 184, 186–194, 201, 215, 220, 230
  Ressel's 81
  developed countries 196–197
social justice 183
social sciences 180, 183, 194
Society for the Preservation of the Wild Fauna of the Empire 170
socio-economic analysis 220
soil 44, 123, 126, 159
  carbon stock 207–208
  chemistry 125
  expectation value 44
  see also Erosion
South Africa 31, 63, 72–73, 103, 112, 118, 126, 136, 170
South America 173, 174
  see also forest of Amazon
South Australia 73, 127–130
Southeast Asia 173
Spain 93, 118, 169
spatial analysis 142
species
  conservation 174–176
  migration 210–211, 214–215
Spiegel-Relaskop 22
spruce 41, 43, 47, 77, 112, 125, 128
  decline 130
  plantation 32, 45, 74
  reaction against 34, 58
stand density 34
stands 19, 135, 138
Statement of Forest Principles 221
statistical analysis 16, 24–25, 37–38, 104, 109, 112, 116, 118, 126, 134–136, 142–143, 231
Stebbing, Edward Percy (1872–1960) 96
Stirling-Maxwell, Sir John (1866–1956) 87–88
Stoate, Theodore Norman 'Bill' (1895–1979) 126–128
stools *see* coppice
structure 36, 77, 161, 164
subsistence 82, 98, 103, 148, 186–188, 192–193, 230
succession 27, 159
sulphur 125, 130
Sundarbans 176
superphosphate 123, 126
survey
  cartographic 15–16
  ecosystems 179
  habitats 174
  plots 21, 23, 174
  quandrats 174
  social 188
  soil 126
  strips 21, 98, 101, 174
  topographical 83–84
  transects 174
  vegetation 87–88, 172
  wildlife 175
  see also measurement
sustainable development 150, 165, 218, 223–224, 233
  see also Management
sustained yield
  critique 134, 140, 220, 225
  nature-based forestry 161–165
  principle 11, 134, 150–151, 218, 220, 225
  see also Management
Svenska Cellulosa Aktiebolaget (Swedish Cellulose Company, SCA) 128
Sweden 32, 114, 128, 135, 152
swidden *see* slash and burn
Switzerland 16, 35, 58, 149, 152, 162–163, 168–169
sycamore 27, 89
Sylva 3–4
  see also Evelyn, John
Syrach Larsen, Carl (1898–1979) 114–115

T

Tanzania 94, 170, 232
Tasmania 152, 160, 165
taungya 102–103

taxation 40, 78, 91, 142
teak 18, 59, 71, 93–94, 99–103, 120
Tele-Relaskop 22
temperature 72, 97
climate change Ch14.
Thailand 103, 194
Theophrastus (372–287 BC) 2
thinning 36–38, 44, 50, 77, 128–129, 135–138
tigers 174, 176–177, 195
see also Project Tiger
timber 28
see also wood
Timor mountain gum 142–144
tire et aire 27–28, 34
tissue propagation 119
traditional knowledge *see* knowledge
Travancore 99
Trieste 78
tropics
countries 74, 101, 216
deforestation 171–174
forests 16, 29, 63, 94, 96–97, 103, 223, 231, 234
measurement 17, 99, 206, 208
people 16, 186, 192
science 31, 38, 104, 211
species 18, 74, 121
Troup, Robert Scott (1874–1939) 31
Tumbarumba flux measurement 212–213
tumpang sari 103
turpentine 86

U

UK *see* Britain
UN Conferences
1972 Stockholm, Human Environment 147
1992 Rio, Environment and Development, 'Earth Summit' 150, 165, 178, 204, 218
1997 Climate Change Kyoto 204
2012 Rio, Sustainable Development, 'Rio+20' 233
UN, Reducing Emissions from Deforestation and Forest Degradation in Developing Countries (REDD) *see* REDD
United Nations Organisation (UN) 107, 201
see also Decades of Developement
United States of America (USA) 93, 114, 150, 158, 173
Pacific Northwest 128, 131, 165
South and Southeast 116, 129, 131, 135–136, 140–144, 155, 216
University
Aberdeen 70
Australian National 189
British Columbia 127
Edinburgh 70, 96, 100
Florida 116, 129
Georgia 135–136, 140
Giessen 45
Idaho 129
Kasetsart 194
Maine 205
North Carolina 116, 129
Oxford 94
Texas A&M 116
Tübingen 54
Washington State 127–128

V

vegetation 175, 131, 195–196, 203, 208, 214
assemblages 174
classification 87–88, 97–98, 172, 174, 179
models 209, 215
see also weed control
vegetative reproduction 114
Verein Naturschutzpark (Nature Conservation Parks Society) 169
Victoria 150, 156
villages and villagers 6, 27–28, 83, 98, 123, 176, 186–187, 223
councils 183, 189, 190–193
Vilmorin, Philippe Andre de (1776–1862) 111, 118

vision
of conservationists/ecologists 168–170, 175
of foresters 1–3, 27, 34, 40, 47–48, 144, 180, 218
see also nature-based forestry, social forestry, sustainable development
volume tables *see* models, volume

W

Wallace, Alfred Russel (1823–1913) 109, 213
walnut, American black 113
Wangenheim, Friedrich Adam Julius von (1749–1800) 68
Warming, Johannes Eugenius Bülow a.k.a Eugen (1841–1921) 88
weed control 29, 36, 77, 101–103, 121, 130–131, 144, 223
Western Australia 73, 91, 124, 126–127
Westoby, Jack C. (1913–1988) 133, 184–185, 191
Weyerhauser 128
wilderness 153, 157, 168–169, 174, 178–179, 180
wildlife 150–151, 156–159, 170–171, 194–195
willows 89, 118
wind damage 29, 32, 36, 77, 210, 232
women 16, 148, 183, 185–188
wood demands 3, 11, 29, 40–41, 72, 77, 185, 215, 223
hardwoods 41, 96, 124–125, 164, 172
industrialisation 40, 60, 116, 123, 133, 142, 165
naval 3, 78, 80, 110
post-war reconstruction 128
softwoods 41, 71
see also fuel wood
wood exports 40, 232
developing countries 133, 184, 193
French 86
imperial 93, 98, 102
wood production objective 60, 149
see also multiple use
wood quality 116
woodland, woods 6, 147, 172
woodlots 187
World Agroforestry Centre 193
World Bank 195
World Forestry Congress 107, 219
fifth 150
eighth 185
World Resources Institute 172
World Wide Fund for Nature/ World Wildlife Fund (WWF) 176, 195, 222
Württemburg 12, 169

X

xylometer 18

Y

yearly annual percentage return 44
yield tables *see* models, growth and yield
Young, Harold E. (1917–1998) 205

Z

Zobel, Bruce John (1920–2011) 116

CPSIA information can be obtained at www.ICGtesting.com
Printed in the USA
LVOW12*1447160714

394633LV00005B/72/P